# BASIC ELECTRONIC SWITCHING FOR TELEPHONE SYSTEMS

# BASIC ELECTRONIC SWITCHING FOR TELEPHONE SYSTEMS

DAVID TALLEY

*Telecommunications Consultant*

HAYDEN BOOK COMPANY, INC.
Rochelle Park, New Jersey

*Library of Congress Cataloging in Publication Data*

Talley, David.
Basic electronic switching for telephone systems.

Includes index.
1. Telephone switching systems, Electronic.
I. Title.
TK6397.T33 621.387′7 75-5590
ISBN 0-8104-5808-X



*Printed in the United States of America*

5 6 7 8 9 PRINTING

80 81 82 83 YEAR

# PREFACE

*Basic Electronic Switching for Telephone Systems* reviews in simple terms the fundamentals of electromechanical switching systems, from the original direct control step-by-step to the common control panel, rotary, and crossbar categories. It details the growth of electronic techniques and solid-state devices leading to the development of electronic telephone switching systems.

The basic design principles and operations of modern electronic telephone switching systems are analyzed with particular reference to those in current use by telephone companies in the United States. The text and its numerous associated drawings are designed for persons without a formal engineering education. Brief explanations are included of the binary numbering system, binary logic principles, and logic circuits to assist the reader in understanding the stored program concept that is utilized in electronic switching systems.

This book has been written to provide executives, engineers, and technicians in the communications and electronics fields, as well as advanced students, with a broad in-depth knowledge of electronic switching techniques and the functions of contemporary electronic telephone central offices. The reader is referred to a previous book of the author, *Basic Telephone Switching Systems,* also published by the Hayden Book Company, for more detailed information on the telephony art, electromechanical switching systems, and interoffice signaling methods.

*New York, N.Y.* David Talley

# CONTENTS

# BASIC ELECTRONIC SWITCHING FOR TELEPHONE SYSTEMS

# 1 TELECOMMUNICATION SWITCHING SYSTEMS AND NETWORKS

## Importance of Switching to Telecommunication Systems

A most important requisite of any telecommunication system, whether used for voice, data, teleprinter, television, or other forms of electrical communications, is its interconnection capability. The system must be able to connect quickly and accurately any station or terminal with any other one in the entire communications network.

These interconnections and switching networks have made possible the worldwide growth of telephone communications. A telephone system is normally comprised of one or more central offices to serve telephone customers in a town, city, or other locality. Several central offices serving the same local calling area are interconnected by a network of interoffice trunks. To reach telephones in other localities, whether in the same or distant states or in other countries of the world, the local central offices are connected through *toll centers* to national and international switching networks.

In the United States, a nationwide intertoll network designated Direct Distance Dialing (DDD) provides the facilities for handling telephone calls nationally and to most countries in the world. For instance, dialing seven digits will connect a telephone to any other one in its local area. Dialing ten digits will permit connections to any one of more than 130 million phones in the United States, Canada, and other countries on the North American continent. Moreover, international dialing has now extended to England and a number of other European countries. Eventually, by dialing eleven to thirteen digits, one will be able to reach almost any telephone in the world. Table 1-1 lists the number of telephones in major areas of the world. It is anticipated that many of them soon will be reached by direct dialing.

## Functions of Central Offices

In a telephone system, the local switching center (usually designated the *end central office*) directly serves the telephones, dataphone sets, Picturephone® units, and other customer equipment connected to it. Calls des-

**Table 1-1.** Telephones in the World

| Country | No. of Telephones in 1974 |
|---|---|
| United States | 138,286,000 |
| Japan | 38,697,901 |
| United Kingdom | 19,095,317 |
| West Germany | 17,802,646 |
| U.S.S.R. | 14,260,700 |
| Italy | 12,611,653 |
| Canada | 11,688,292 |
| France | 11,337,000 |
| World | 336,297,000 |

From "The World's Telephones," AT & T Co., 1974.

tined for outside its service area are usually routed by the end central office to an associated toll switching center. This toll center will have access to the intertoll or DDD network of the Bell System and to international and global networks as later explained. Figure 1-1 diagrams the switching arrangements between central offices and a toll center.

The local or end central office contains various equipment for handling originating calls from customers and, likewise, for completing incoming calls to its telephone stations. It directly receives the digits dialed by the call-originating station, translates them, performs the required switching actions, and transmits details of the called number to the distant central office or toll center as may be required. End central offices also furnish transmission and signaling facilities including ringing, charging for calls, maintenance, and testing features. The application of electronic switching to such central offices is the subject of this book.

Toll central offices are composed of toll centers and primary, sectional, and regional centers. These switching centers, as subsequently explained, interconnect trunks for routing calls to their destinations but have no direct access to telephone stations. In handling a toll or long-distance call, the originating central office transmits the ten-digit number of the called party (three-digit area code plus seven-digit directory number) to its associated toll center. The first three digits (area code) or the first six digits (area code plus central office code) of the called number are translated by the toll center, depending on the call's destination, routes available, and other factors. The toll switching center then performs the necessary switching functions to route the call toward its destination.

## Telephone Numbering Plans

To provide for local, national, and international or global communications, it is essential that every telephone station be uniquely identified, as

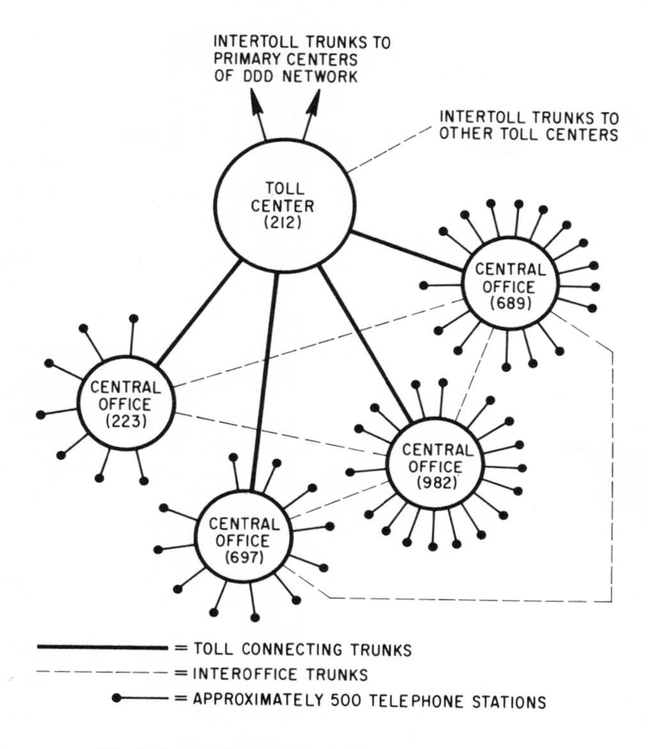


**Fig. 1-1.** Local Telephone Switching Network

are social security numbers. Consequently, a seven-digit telephone number has been standardized in the United States and many other countries. The first three digits make up the central office code and the last four contain the line number. For example, 555 is the central office designation or office code of telephone number 555-1212. On the basis of a four-digit line number, each central office will have a theoretical numbering capacity of 10,000 lines in the sequence 0000 to 9999. (It may be of interest to know that about 27,300 central offices were required to serve the approximately 132 million telephones in the United States in 1974, including those of the Bell System, the General System, and Independent Telephone companies.)

It is believed that the telephone numbering plan started with the manual central office switchboard. The largest switchboard had a capacity of 10,000 line jacks which were numbered from 0000 to 9999. In cities and other localities where additional central offices were needed, each four-digit line number was preceded by the designation of its central office, usually named after the locality or a historic personality.

With the advent of dial automatic switching, this numbering scheme initially was continued in the larger cities, except that only the first three letters of the central office name were dialed before the line number. In the smaller towns and cities, one or two digits were originally used as the office code because the electromechanical switching equipment was of the step-by-step type and required a separate level for each digit dialed.

The rapid growth of telephone service, particularly in urban areas, began to deplete the number of office codes derived from the first three letters of the central office name. Moreover, only 330 codes were found usable from the possible 1,000 (10 × 10 × 10) combinations of the three digits. As a result, the assignment of two letters, and a numeral in place of the third letter, was next introduced. This plan increased to 576 the total number of usable central office codes. The continued demand for telephones, however, led to the present all-number calling plan (ANC) whereby numbers replaced all letter combinations for office codes. A total of 800 (8 × 10 × 10) possible combinations is provided by ANC because the numerals 1 and 0 are not usable for the first digit of an office code. Because of the requirements for geographical area codes and special-purpose codes, such as 411 for information and 911, the national emergency code for police assistance, only 598 central office codes are available for assignment. Considering a maximum of 10,000 numbers per central office code, it is possible to have 5,980,000 telephone numbers in each area code region.

For long-distance dialing purposes utilizing the DDD network, the Bell System, with the cooperation of other telephone companies, has divided the United States and Canada into approximately 125 telephone regions, each assigned a three-figure area code. The specific area code is dialed only if the called telephone number has an area different from that of the calling telephone number.

The three-digit area code is added to the assigned seven-digit directory number to make a total of ten digits to be dialed. The first digit of the area code does *not* contain the numbers 0 or 1. The second or middle digit *must* be either 0 or 1, and the last digit utilizes *all* numerals from 1 to 9 and 0. This

**Table 1-2.** Area Code Numbering Plan

| Location | Area Code |
|---|---|
| Boston | 617 |
| New York | 212 |
| Miami | 305 |
| Chicago | 312 |
| Washington, D.C. | 202 |
| Dallas | 214 |
| Los Angeles | 213 |
| San Francisco | 415 |
| San Juan, P.R. | 809 |
| Montreal, Canada | 514 |
| Mexico City | 903 |

area code numbering plan provides a total of 160 (8 × 2 × 10) possible codes. The 0 and 1 designations for the middle digit enable the central office switching equipment to distinguish a toll or long-distance call from a local area call. Some typical area codes are listed in Table 1-2. Area codes are used to route calls from one telephone switching network area to another but are not used within the same network area.

## Direct Distance Dialing Network

The Bell System's Direct Distance Dialing (DDD) network has made possible the interconnection of telephone stations in the United States, Canada, Mexico, and almost every country on the North American continent. This network is controlled by switching centers designated, in order of importance, *regional* (Class 1), *sectional* (Class 2), and *primary* (Class 3) centers. These three centers are connected to the *toll centers* (Class 4 offices) referred to in Fig. 1-1. A simplified representation of the DDD network is illustrated in Fig. 1-2.

*End central offices* (Class 5 offices) connect to the DDD network through the toll centers. Calls between local or end central offices (Class 5) in the same area code region may be routed over interoffice trunks or through a tandem office as shown in Fig. 1-2. Calls through toll centers normally are sent over direct high-usage routes. If these routes are busy or not available, alternate switching routes through primary, sectional, and regional switching centers are selected in that sequence.

The aforementioned switching centers in the DDD network are interconnected by coaxial cables and point-to-point microwave radio systems. To provide the many thousands of voice and data channels, carrier transmission of signals is used on these coaxial cable and microwave radio facilities.

Referring to Fig. 1-2, let us follow the routing of a telephone call from Boston to Los Angeles. Dialing area code 213, followed by the seven digits of the called party in Los Angeles, will cause the originating central office (Class 5) to switch the call to its associated toll center (Class 4). The switching equipment in this toll center, upon receiving the 213 area code, will immediately pass the call to the primary center serving the Boston area. The primary center (Class 3) first will try to complete the call over the high-usage trunks designated 1, 2, and 3 in Fig. 1-2. If routes 1 and 2 are busy, the call will be sent over route 3 to the sectional center (Class 2) in Los Angeles. This Class 2 office, consequently, will switch the call via final route trunks through the primary (Class 3) and toll center (Class 4) offices serving the Los Angeles area to reach the particular end central office (Class 5) serving the called party.

If the outgoing intertoll trunks from the Boston primary center (Class 3) to Los Angeles are busy, route 4 to the sectional center will be selected. The switching equipment in this Class 2 office will attempt to complete the call over routes marked A, B, or C before selecting final route D to the re-

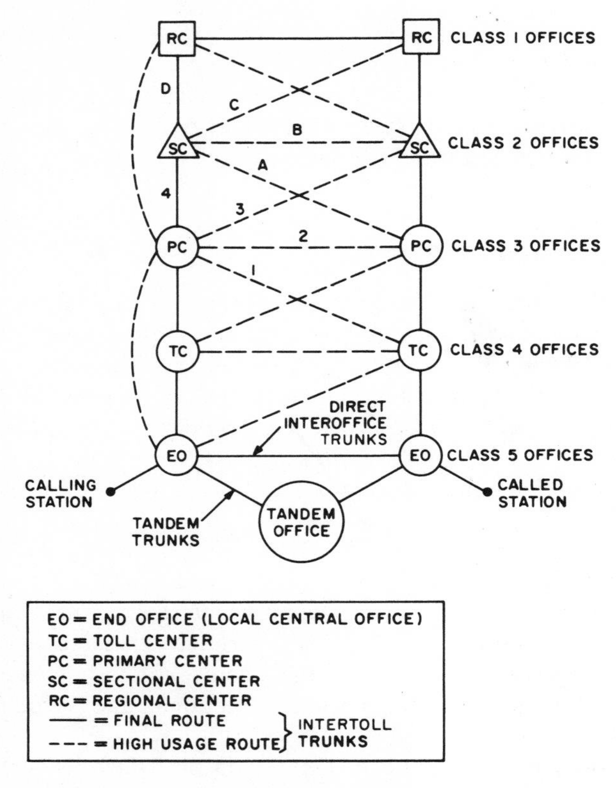


**Fig. 1-2.** Simplified Direct-Distance-Dialing Network

gional center (Class 1). Regional centers are interconnected by direct final route trunks. Therefore, the call can be routed from the Los Angeles regional center (Class 1) to the desired end central office (Class 5) through the respective sectional, primary, and toll centers shown in Fig. 1-2.

## Worldwide Dialing

Worldwide dialing of telephone calls and other communication interconnections has been substantially advanced by recent developments in the communications field. These improvements include the installation of high-capacity and repeater-type telephone cables under the oceans, and wide-band communication satellites in synchronous orbits around the earth. These new facilities provide many thousands of channels for voice, data, and related communication needs.

To facilitate the use of these global communication systems, a world zone-numbering plan has been formulated by the International Telegraph and Telephone Consultative Committee (CCITT) of the International Tele-

**Table 1-3.** World Telephone Numbering Plan

| World Telephone Zone Code | Zone or Region |
|---|---|
| 1 | North America including Hawaii, Central America, and Caribbean area except Cuba |
| 2 | Africa |
| 3 | Europe |
| 4 | Europe |
| 5 | South America and Cuba |
| 6 | Australia and South Pacific |
| 7 | U.S.S.R. |
| 8 | North Pacific including Eastern Asia |
| 9 | Far East and Middle East |

communications Union. This plan divides the world into nine telephone numbering zones as illustrated in Table 1-3.

A telephone user in any country would be connected to the world switching network after first dialing an access code of two or three digits. The user would next dial the desired world zone number of the called party. For example, on a call to Sydney, Australia, the number 6 would be dialed followed by a two-digit regional code, and then the complete telephone number of the called person in Sydney, Australia. In the United States, Canada, and other countries of North America, the area codes correspond to the country or regional codes. Dialing the two- or three-digit access code to the world network will thus make it possible in the near future to connect to almost any telephone in the world after dialing a maximum of eleven to thirteen digits.

## Telephone Station Equipment

The telephone instrument (commonly called the *subset*), used at customers' stations to initiate or answer calls, is a sophisticated and reliable device designed to function with different switching systems. It normally consists of a handset containing the receiver and transmitter units (some types may include a rotary or pushbutton dial), and a small apparatus box containing the ringer, transmission circuit, rotary dial, or pushbutton keyset, and the hookswitch for the handset. Typical telephone instruments are shown in Figs. 1-3 and 1-5. The telephone subset performs these general functions:

1. Signals an incoming call by the bell-ringing signal
2. Signals the central office equipment whenever a call is originated or answered (handset off-hook), and terminated upon disconnection (handset on-hook)
3. Transmits dial pulses or pushbutton tones (address signals) to the central office equipment

4. Enables two-way operation (transmission and reception) over the two-wire telephone line
5. Converts speech to electrical signals and vice versa through the handset.

Telephone subsets also may be designed for four-wire operation which requires separate pairs of wires for each direction of transmission. Because of economic considerations, a single pair of conductors is employed to connect telephone stations to the central office on a two-wire transmission basis. Four-wire circuits, however, are normally employed for toll and long-distance trunks and in the nationwide DDD network. Likewise, interoffice and short-haul trunks, which utilize carrier or multiplex facilities, use separate pairs for each transmission direction to provide four-wire operation.

## Rotary Dial and Pushbutton Dialing

The previous references to dialing of telephone numbers include the use of either the rotary dial or pushbutton dialing by the telephone user. Figure 1-3 is a photograph of a typical telephone instrument or subset equipped with a rotary dial. The rotary dial, which initially was developed in the early days of automatic switching equipment, is essentially a circular device with ten finger-holes numbered 1 to 9 and 0, arranged in a counterclockwise direction around its periphery. The rotary dial opens and closes contacts as it spins back to its normal stop position, from the digit to which it was turned, thereby generating *direct current* (dc) pulses. These pulses, which correspond to the numeral of the selected digit, are used as "address" signals for instructing central office switching equipment regarding the called telephone number.

Rotary dials generate pulses at the rate of 10 per second irrespective of the digit dialed. Dials on switchboards of private branch exchanges (PBX) or private automatic branch exchanges (PABX) can send 20 pulses per second, thus reducing the operating time of handling calls.

Some 12 or more seconds are normally required to dial the seven digits of a local telephone number, and at least 14 seconds to dial ten digits when making toll or long-distance calls. This time period has proven satisfactory for electromechanical switching offices, particularly the step-by-step type which function directly from dial pulses as sent from the telephone dial. The aforementioned dial-time periods mean that certain common equipment in the central office, such as the register-sender, will not function until all the dial pulses are received. It is apparent, therefore, that the longer the dialing period, the greater the amount of common control and related switching equipment that will be needed to handle the many simultaneous calls during busy-hour periods.

The very high switching speeds made possible by electronic switching systems have stressed the need of more rapid and accurate dialing means by telephone users. Key pulsing, a form of multifrequency (MF) pushbutton

**Fig. 1-3.** Telephone with Rotary Dial (Courtesy of AT&T Co.)

calling using combinations of two tones, had been used for many years by operators on toll and dial service assistance switchboards. This particular multi-frequency dialing system is not suited for use with telephone instruments because of the need to provide filter circuits to guard against voice interference and other considerations. As a result, a new pushbutton dialing method designated *Touch Tone®* has been developed by the Bell Telephone Laboratories for use with electronic switching systems. It is also usable with present electromechanical central offices by providing suitable tone-receiving and conversion equipment. Touch Tone® dialing, also known as touch calling, utilizes eight frequencies in the 700-1700 Hz range divided into low-band and high-band tones, as illustrated in Fig. 1-4. These eight frequencies, selected to avoid harmonically related interference from voice signals, are designed for a maximum of sixteen pushbuttons containing four rows and four columns on the faceplate of the pushbutton dial. A typical pushbutton or touch calling telephone set in current use is shown in Fig. 1-5. The twelve pushbuttons are arranged in four rows and three columns as illustrated. Pushbuttons designated * and #, corresponding to digits 11 and 12, are primarily for use with electronic switching offices. The functional circuit of the GTE Automatic Electric touch calling keyset is represented by the schematic diagram of Fig. 1-6. Transistor Q1 and its associated components—inductors L-3 and L-4, capacitor C-3, resistors R-2 and R-3, and diode D-1—make up an *audio*

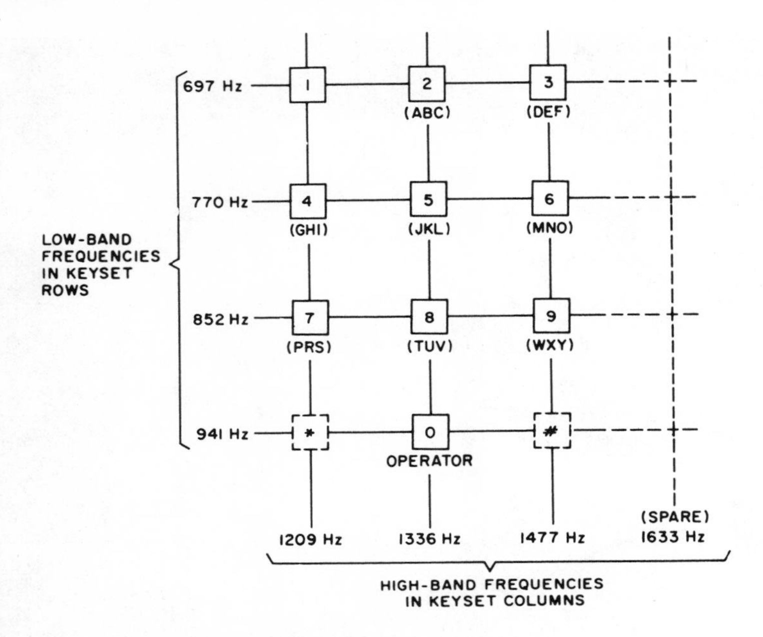


**Fig. 1-4.** Pushbutton Dialing Frequencies

*oscillator.* The audio frequencies produced by this circuit are controlled by the operation of the keyset pushbuttons in accordance with the frequency plan outlined in Fig. 1-4.

As an example of the keyset's operation, assume that digit 7 (PRS), located in the third row of the first column, is operated. Referring to Fig. 1-6, this action will close row 3 contacts of switch S-1 and capacitor C-1 will be connected to the third tap on inductor L-1, thereby establishing a resonant

**Fig. 1-5.** Pushbutton Telephone (Courtesy of GTE Automatic Electric Co.)

circuit for producing the 852 Hz tone. At the same time, another resonant circuit for creating the 1209 Hz tone will be constituted by the closure of the S-2 switch contacts in column 1 to connect capacitor C-2 to the first tap on inductor L-2. Switch S-3, which is common to both the S-1 and S-2 switch contacts, cannot operate until the pushbutton has been fully depressed. When the contacts of common switch S-3 are closed, a low dc voltage (produced from the diode rectifier unit bridged across the telephone line) will be connected to activate transistor Q1 and cause the generation of the 852 and 1209 Hz audio tones. RV-1 and RV-2 are semiconductor rectifiers, termed varistors, which are employed to reduce interactions between the respective adjacent frequencies. The receiver (REC.) in the handset will be shunted by the 5100 ohm resistor R-1 whenever switch S-3 is operated, but the calling party will be able to hear the transmitted tone signals corresponding to the dialed or keyed digits, although at a reduced level.

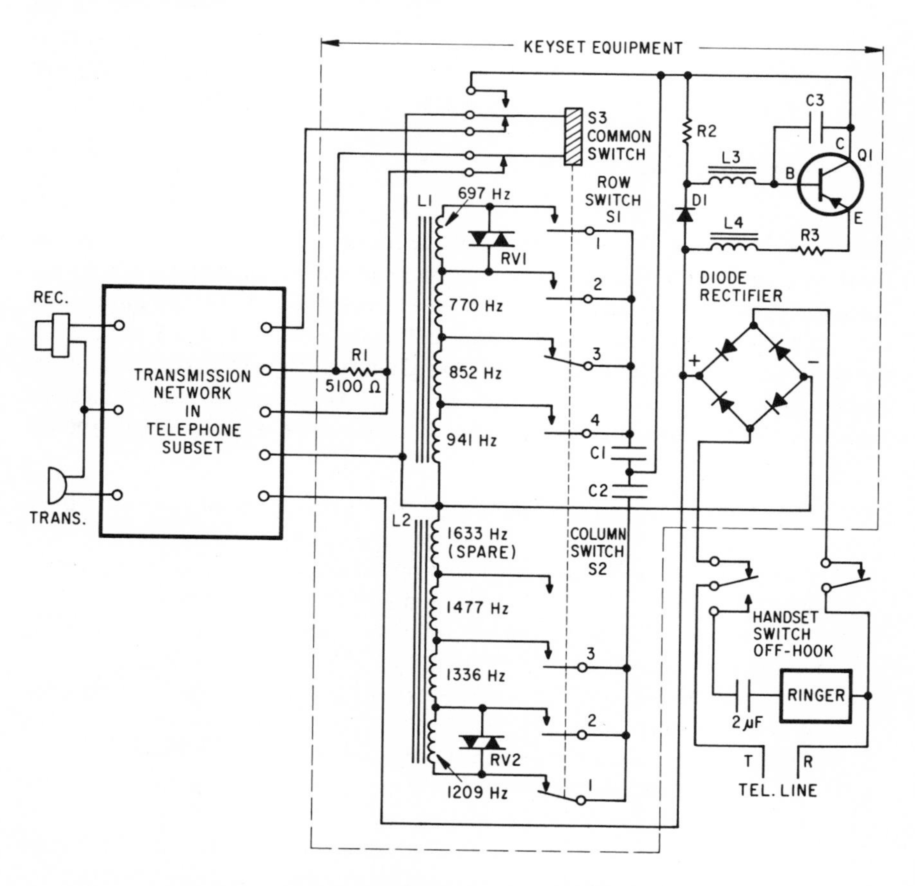


**Fig. 1-6.** Schematic Diagram of Touch Calling Keyset

An important advantage of pushbutton or Touch Tone® dialing is that the audio tones generated for each digit are in the voice-frequency range. Therefore, they can be transmitted anywhere over the DDD switching network in the same way as speech signals. This outstanding characteristic can be utilized to provide many new services for telephone customers. For instance, a customer may dial the number of an automatic department store, and place an order by dialing the digits corresponding to his or her account number and to a coded list of desired items. These services are not possible with the rotary dial because the dc pulses normally produced control only the equipment in the local central office. Furthermore, almost all interoffice and toll trunks employ carrier transmission so that metallic paths are not available for handling dc dial pulses.

Additional equipment is required in electromechanical central offices to receive and translate tone signals from pushbutton dials. For step-by-step offices, the touch calling signals are converted into direct current dial pulses to permit direct operation of the selector switches. In common control offices, such as the No. 5 Crossbar, the Touch Tone® signals are detected and translated into the proper form required by the given switching equipment.

## Automatic Switching Systems

Before discussing electronic switching systems, it is desirable to understand the concepts involved in the development of automatic telephone switching, and the operations of major electromechanical switching systems. Automatic switching equipment initially was designed to aid and later to replace a telephone operator and a manual switchboard. The evolvement of electromechanical systems has simulated to a large degree the switching functions of the cords, plugs, and jacks on manual switchboards. These functions performed by the operator in processing a local call on a manual switchboard are as follows:

1. The operator scans the switchboard panels for lights indicating an originating call.
2. When a lamp lights, the operator immediately notes its location, the identity of the calling line's jack, and records this information.
3. The operator inserts the plug of an answering cord into the jack of the calling line and announces, "Operator."
4. The called line number is received from the calling party and recorded by the operator.
5. The jack of the called party on the switchboard is located.
6. The operator determines the state of the called line by testing the sleeve of its jack with the tip of her calling cord's plug.
7. If the called line tests idle (click is not heard by the operator), the calling cord's plug is inserted into the called line's jack to complete the con-

nection. Ringing is started automatically by the relay equipment in the cord circuit.

8. If the called line tests busy (click is heard by the operator) the plug of the calling cord is inserted into a busy-tone jack which transmits a busy tone to the calling party.
9. When the connection is completed, per steps 7 or 8 above, the operator is available to handle another call.
10. The progress of the call is indicated by lamps on the keyshelf of the switchboard which are associated with each answering and calling cord.
11. The operator releases the connection by removing the answering and calling cords when both lamps light to indicate that the calling and called parties have disconnected. The operator can perform this function while processing other calls.

Automatic or electromechanical switching equipments do not *exactly* duplicate the direct interconnections completed by the operator using the cords, jacks, keys, and other switchboard equipment. Relays, vertical and rotary motions of stepping switches, brush-type selectors, and crosspoint contacts of crossbar switches are utilized for interconnecting telephone lines in electromechanical central offices. Note that central offices, whether manual, electromechanical, or electronic switching types, are not designed for the simultaneous connections of all or even a majority of their customers. For instance, to permit all customers to talk to each other at the same time, the number of required interconnections in the central office would amount to one-half the number of customers. As a rule, only a small portion of customers, normally 10 percent to 14 percent, simultaneously originates calls within a given time of busy-hour periods. Consequently, the central office equipment needs the capacity to serve only that percent of the total lines. In a central office serving 6,000 lines, for example, approximately 700 calls would usually be connected or in progress of completion at any one instant in busy-hour periods. Thus, only the amount of equipment capable of handling that number of simultaneous connections need be provided. This principle is also followed in the design of electronic switching central offices.

## Direct Control Step-by-Step Switching Systems

Electromechanical switching systems are usually divided into two general categories: *step-by-step* (S × S) *direct control* and *common control* systems. The original direct control switching method evolved from the invention of the Strowger switch used in the initial automatic dial system. It would advance step-by-step in unison with the dial pulses. The S × S method may be defined as a direct progressive control system in which stepping-type switches are directly controlled by pulses from customer's dials. As a result, there is an established correlation between the speed of the dial pulses and the motion of the stepping switches in the central office. The S × S switching technique is still used today in the majority of electromechanical central offices.

Figure 1-7 is a block schematic drawing which shows the principal elements of the switch train in a S × S central office and their relations with the dialed digits of a four-digit line number. The stepping or selection switches are designed for both vertical and horizontal movements which are called, respectively, the selection and hunting stages. These switches can step a maximum of ten levels vertically in unison with dial pulses and then, in the hunting stage, automatically rotate horizontally over ten terminals to search for an idle path to the next selector or other circuit. The time interval between the digits dialed by customers is the factor limiting to ten the number of horizontal terminals provided in the hunting stage. The last selecting switch is called the connector because it connects directly to the terminals of the called line number. The connector switch has two selection stages but does not hunt. The vertical stepping is controlled by the tens digit and the horizontal rotation is controlled by the units digit of the called line number.

To illustrate the processing of a local or intraoffice call within a small S × S central office, let us assume that only four-digit line numbers are assigned, and that the number 4356 is to be called from a customer's telephone. As we see in Fig. 1-7, when the calling party initiates the call (handset off-hook), an idle linefinder switch (always associated with a first selector-finds and connects to the calling line. A dial tone is sent to the calling party by this first selector. Dialing the thousands digit 4 steps the first selector vertically to its fourth level. Then the selector rotates horizontally to select an idle path to a second selector that serves the 4000 number series. The next digit, 3, will step the second selector to its third level where it will hunt for an idle path to connector switches associated with lines in the 4300–4399 number sequence. The next or tens digit 5 will cause the connector switch to move to its fifth level. Since this switch has no hunting feature, the last or units digit 6 will step this connector horizontally six positions, thereby connecting to the called line

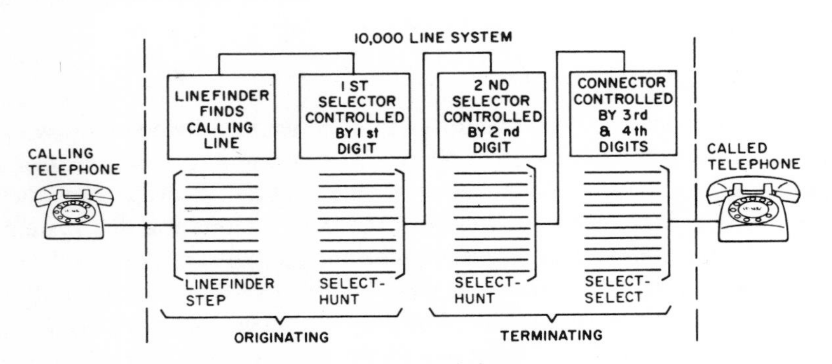


**Fig. 1-7.** Direct Control S x S Switching System

number 4356. The connector next tests the called line; if it is idle, ringing current is applied. If the line is busy, a busy tone is returned to the caller.

Direct control switching systems, such as the step-by-step, have a number of disadvantages, however, especially when several central offices are required in the same local service area. A prime drawback is that the office code digits define the exact levels to which selectors are stepped. For example, office code 637 will cause the first selector to step to the sixth level, the second selector to the third level, and the third selector to its seventh level. It is not possible to change this fixed relationship between central office codes and locations on selector switches of interconnecting paths or outgoing trunks to other central offices. Once the customer has dialed the three digits of an office code, it is not feasible to select an alternate route if all paths or trunks to the called central office should be busy. In such cases, a busy signal will be sent back to the calling party.

Another disadvantage is that the tens and units digits of the called line number specifically determine its location on the connector switch. Traffic overloads, therefore, could arise if numerous customers who receive many calls are assigned line numbers within the same hundreds group, for example: 3100–3199, 6700–6799, 8000–8099, etc.

## Common Control Switching Systems

For maximum efficiency, switching systems theoretically should have full access and nonblocking capabilities. *Full access,* however, actually means that every telephone line can be connected to every other telephone but not necessarily at the same time. *Nonblocking* means that any originating call in a central office can be connected to any other telephone irrespective of the number of existing connections. These goals are substantially obtained in most common control switching systems.

Common control switching systems of the electromechanical type in current use include the panel and crossbar categories manufactured by Western Electric Company for the Bell System, and the rotary and director systems which are used by some General System and Independent Telephone companies. The director system, which was developed by GTE Automatic Electric Company, applies some of the main features of common control operations to step-by-step central offices. This common control switching method is used by a number of General System telephone companies.

Two main characteristics of common control switching systems are the employment of large capacity switches or switch frames, and use of groups of common control circuits commonly termed ***register-senders, originating senders,*** or ***senders,*** for temporary storage and translation of received dial pulses into appropriate signals for controlling the switching network. For instance, the switch frames in panel dial central offices have access to 500 levels. In the crossbar system, a combination of crossbar switches can provide access to several thousand outlets. On the other hand, selectors in S × S offices have a capacity of only ten levels, with ten paths on each level, or a possible 100 outlets.

Moreover, in common control offices, the switching operations are separated from the actions controlling them. This permits a particular group of common control circuits, such as register-senders, to direct the routing of connections through the switching network for many calls at the same time. Furthermore, as soon as a particular register-sender completes the routing of a call to its destination, it releases and becomes immediately available to handle other calls.

## Panel Dial System

The panel dial system, which initially was developed in the 1920s by the Bell System to replace manual central offices in the larger cities, is a common control, progressive switching type. It has five major categories of vertical selector switching frames: *linefinder, district,* and *office* for handling outgoing calls; and *incoming* and *final* frames for completing connections to the called line number. Two principal common-control devices are also utilized, namely, the *register-sender* and the *decoder* or *decoder-marker* circuits. The register-sender provides a dial tone to the calling line, temporarily stores the called line number as dialed by the calling party, and connects to a decoder circuit. The decoder translates the first three digits (area or office code) into proper routing instructions which are sent back to the register-sender. The register-sender then proceeds to control the selection operations of the switching network, consisting of the district, office, incoming, and final frames. A brief description of the equipment used in the panel system and its call-processing operations follows. (Portions of the balance of this chapter are adapted from the author's *Basic Telephone Switching Systems,* published by Hayden Book Company in 1969.)

Linefinder frames usually are equipped with ten vertical banks of forty line terminals per bank or 400 per frame. The other selector frames have five vertical banks, designated 0 to 4 inclusive, each having 100 sets of three terminals corresponding to the standard tip, ring, and sleeve conductor terminology. These terminals protrude through both sides of the banks for contacting the brushes on selectors. Each selector frame has a capacity of sixty vertical selectors which are driven by cork rollers, one for updrive and the other for downdrive or restore to normal. These cork rollers are continuously rotated by electric motors, two on each selector frame. An eighteen-position rotary switch, called a sequence switch, both is associated with and controls the operation of each selector on the frame. This switch can be rotated to different positions by a revolving vertical drive shaft, operated by the earlier-mentioned electric motor, under control of an electromagnetic clutch and related relay circuits. Various circuit combinations are made or changed as the

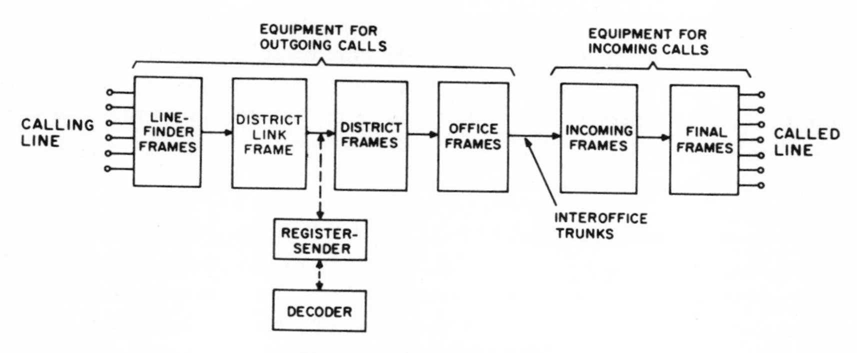


**Fig. 1-8.** Panel Dial System

sequence switch moves to different positions, thus controlling the diverse functions in handling a call.

Figure 1-8 is a simplified block diagram of a panel switching system. Let us follow a local interoffice call through the switching network. When the calling party lifts the handset, a linefinder is used to search for the calling line's terminals on a particular bank of the linefinder frame. At the same time, a district-sender link circuit functions to select an idle register-sender which will send a dial tone to the calling telephone. Assuming that the number 637–7890 is dialed, as soon as the first three digits (637) are received by the register-sender, it will connect to a decoder and transmit the digits 637. The decoder translates this office code into the particular brush and group selections to be made by the district and office selectors involved in the call, returns this information to the register-sender, and immediately releases. The register-sender now proceeds to direct the district selector to a specific bank and group of terminals to find an idle path to an office selector. It then directs the chosen office selector to the particular bank and group location of the outgoing trunks to central office 637. The office selector hunts for an idle trunk in this group of terminals, thereby completing the routing instructions that had been received by the register-sender from the decoder.

The interoffice trunk selected by the office selector terminates on a selector circuit in an incoming frame of panel office 637. The register-sender, having previously recorded 7890 as the called line number, directs the incoming selector to trip its brush 3 and then advance to group 3, the top group of twenty-five terminals on bank 3 of the incoming frame. This particular group of bank terminals has access only to the final frame selectors which can connect to line numbers in the 7500–7999 series. The register-sender next commands the final selector to trip its brush 3 and then travel to terminal 90, thereby connecting to line number 7890, which is assumed to be idle. This completes the functions of the register-sender and it releases. Simultaneously, the district selector circuit cuts through the talking path from the linefinder to the interoffice trunk, and the incoming selector circuit applies a ringing current to the called line. If the called line had tested busy, the final selector would have returned to normal and a busy tone would have been returned to the calling party. When the calling party hangs up, the district, office, incoming, and final selectors disconnect and restore to normal in that order.

## Rotary Switching System

The same design principles of the panel system are followed in the rotary switching system, except that different types of power-driven switch frames are employed. The rotary system, as its name implies, utilizes rotary selectors on semicircular banks for switching operations. It includes similar power-driven sequence switches and common control equipment, such as register-senders, decoders or translators, and linefinders, as used in the panel system.

Trial installations of both the rotary and panel systems were originally made in the early 1920s, but subsequently, the Bell System decided to standardize the panel system for the large cities in the United States. The rotary system has been substantially improved over the years, especially by the Bell Manufacturing Co. of Antwerp, Belgium, which is affiliated with the International Telephone & Telegraph Corp. Installations of the types 7-A to 7-E Rotary System have been made in many countries of Europe and other parts of the world. Some rotary systems have been installed for Independent Telephone companies in the United States.

## Common Control Step-by-Step System

The rapid growth of telephone requirements in small communities, the advent of the Direct Distance Dialing network, and extended area service (EAS) have necessitated the assignment of standard seven-digit numbers to all customer telephones served by step-by-step (S × S) central offices. These factors and the need to provide alternate routings for outgoing traffic in S × S offices, accentuated the necessity to apply common control operations to existing S × S offices. The techniques for this purpose had been known for many years, but had not been previously considered economical or essential for small S × S central offices.

One method for incorporating major common control features into existing S × S installations without modifying the selector switches and switch train is the 101 Director System developed by GTE Automatic Electric Co. It provides register-senders with associated access relay equipment, and an electronic translator which may be common to as many as 100 register-senders. This translator is a very high-speed electronic device which can receive and translate a maximum of six digits (area and office codes), and provide routing instructions to the register-sender for completing the call. A block diagram showing these common control techniques as applied to a typical S × S office is presented in Fig. 1-9. Note that the access circuit for selecting a register-sender is inserted between the linefinder and first selector in the switch train.

On local and intraoffice calls, the first two digits of the office code usually must be absorbed because only the third digit is required to direct the first selector to the proper level for selecting an idle path to the next selector in the switch train. For example, assume the number 637−4321 is dialed. The register-sender immediately repeats the first digit 6 to the first selector, and also to an idle electronic translator. The translator, recognizing that the first digit 6 may signify a local or intraoffice call, orders the register-sender to release the first selector. When the second digit 3 is dialed, the code 63 is sent to the translator which, knowing it as the elements of a local call, directs the register-sender to absorb these two digits, and to repeat the next dialed digit to the first selector. Therefore, as soon as the third digit 7 is dialed, the register-sender repeats it to the first selector, as was ordered by the translator. At the same time, the complete office code digits 637 are sent by the register-sender to the translator, as an indication of completion of office code dialing. The translator will now cause the release of the register-sender. The subsequent dialing of the called line number 4321 will directly control the operation of the other selectors in the switch train in the usual manner.

On interoffice calls when all office code digits are required for routing the call, the register-sender can provide alternate routings. For instance, assume that all ten trunks to a certain office, which appear on level 4, second selector switch, are busy. The register-sender, upon receiving this indication, and knowing that an alternate route is required in accordance with instruc-

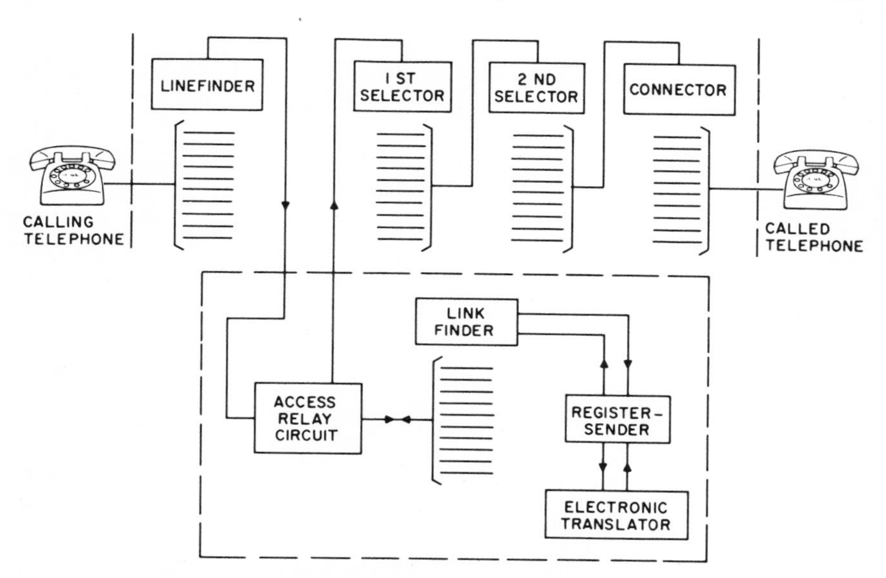


**Fig. 1-9.** Common Control Applied to Step-by-Step Switching System

tions previously received from the translator, will release the first and second selectors by momentarily opening the dialing loop (tip and ring leads). Then, the register-sender will direct the first and second selectors, in sequence, to other levels in accordance with instructions received from the translator, in order to select an idle trunk in the alternate route.

## Crossbar Switching Systems

The panel dial and rotary switching common control systems have numerous disadvantages. Among them are the use of power-driven selectors and sequence switches, sliding contacts on bank terminals which are a source of noise, and the comparatively long time required by selectors to find idle paths and to complete calls. Moreover, considerable equipment, including sequence switches and relays directly associated with each selector, are used for only a few seconds during the establishment and disconnection periods of a call. As a result, substantial quantities of apparatus are needed in such common control central offices to handle normal traffic volume. These factors and the high cost of power-driven mechanisms led to the development by the Bell Telephone Laboratories of the crossbar switching system utilizing the crossbar switch.

The original crossbar switching design was the No. 1 Crossbar System which the Bell System installed during the 1930s in New York and other large cities. The switching network of the crossbar system is not bound in any manner by the called number assignments. Customer line numbers have only a single appearance on a line-link frame, which handles both originating and terminating calls, in contrast to connecting to both a linefinder and final frame in the panel system. The switching network employs crossbar switches, each normally having 200 sets of crosspoint contacts in twenty horizontal and ten vertical rows on the bank. Each crosspoint actually has twin precious-metal contacts which may be pressed together as in the operation of a relay, thus minimizing sliding contacts.

The principal control device of the crossbar switching system is the *marker circuit* which, in addition to translating dialed digits as the decoder does in the panel system, locates idle paths and trunks for routing calls through the switching network. If first-choice paths through the network are busy, the marker makes a second attempt. Similarly, the marker will select an alternate route if all direct trunks to the called central office are busy. The marker also determines whether the called line number is idle or busy. Consequently, the marker may be compared to a computer because it receives information, refers to its memory for programmed instructions, and then initiates or directs other equipment to take appropriate actions in processing and completing calls.

The simplified block diagram in Fig. 1-10 illustrates the basic elements of the No. 1 Crossbar System. Separate markers and switching networks are

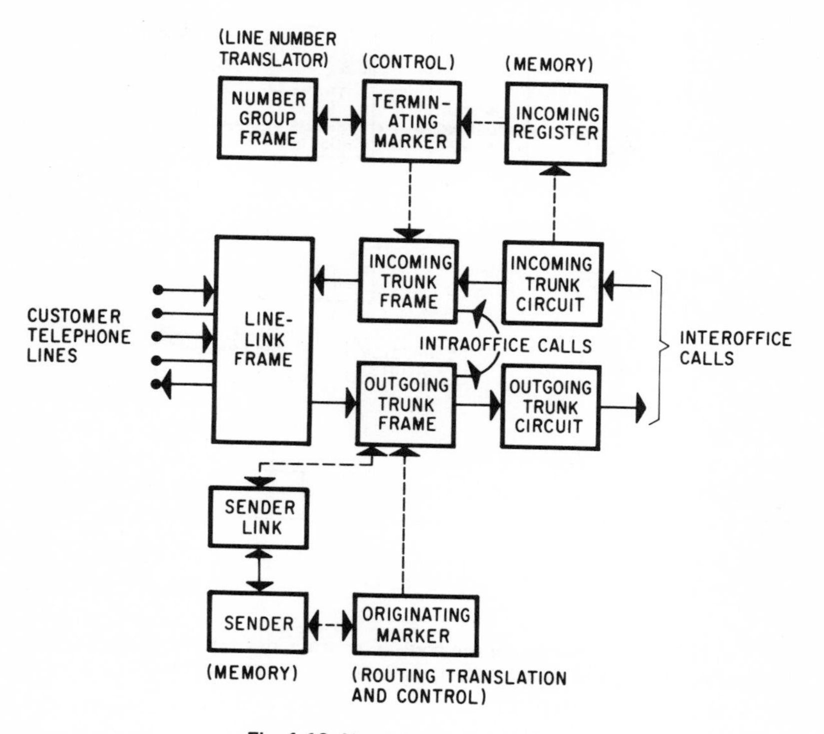


**Fig. 1-10.** No. 1 Crossbar System

employed for originating and terminating traffic, respectively. The number group frame indicated in Fig. 1-10 tells the terminating marker on which particular line-link frame and at what specific crossbar switch crosspoint it will find the line terminal corresponding to the called directory number. The line number on the line-link frame need not necessarily correspond to the directory number. This permits greater flexibility in handling customer traffic loads because, for instance, it is not necessary that all private branch exchange (PBX) lines have consecutive directory numbers as in other electromechanical switching systems.

The No. 5 Crossbar System, which was initially placed in service by the Bell System in 1947, is probably the ultimate in sophisticated electromechanical switching systems. Unlike the No. 1 type, it utilizes one switching network and a single marker to handle both originating and terminating traffic. Figure 1-11 is a block diagram showing the major components of the No. 5 Crossbar System, including the combined switching network (line-link and trunk-link frames) and the marker with its associated register and sender circuits. The connector circuits in this diagram are used to connect the marker with other common control equipments, and to the line-link and trunk-link frames. These connecting paths are utilized only for about 0.2 sec-

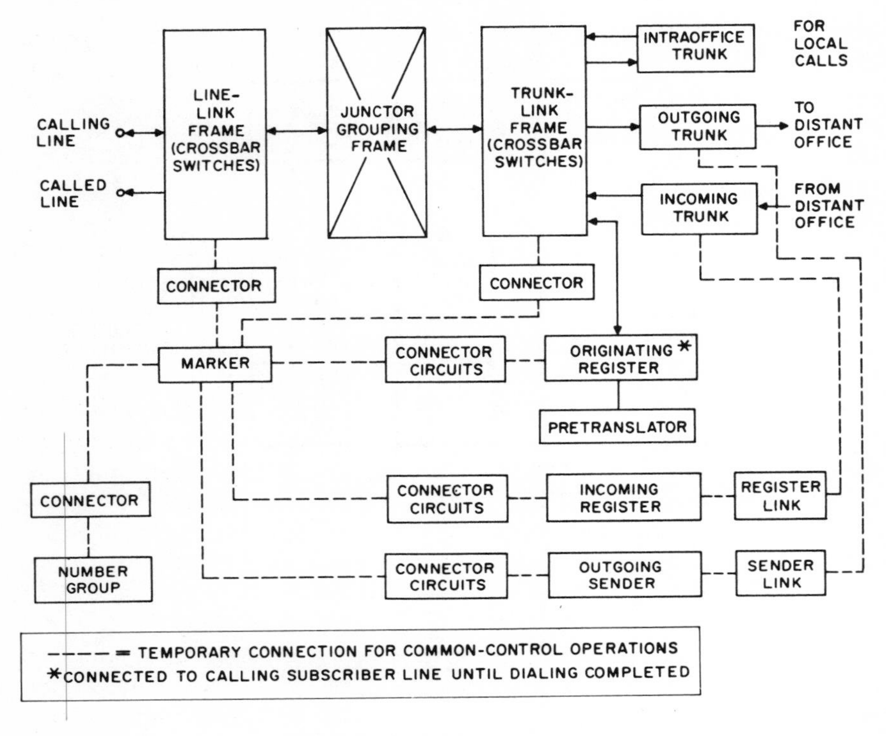


**Fig. 1-11.** No. 5 Crossbar System

ond during the progress of a call. A synopsis of the processes involved in handling an interoffice local call follows.

A customer lifting the handset to initiate a call will cause the operation of the line (L) relay on a particular line-link frame, which in turn activates the associated connector circuit to seize an idle marker. The marker proceeds to locate the calling line on the particular line-link frame, and secures an idle originating register on a trunk-link frame. It then closes a selected path from the calling line terminals on the line-link frame, through the junctor group and trunk-link frames, to the originating sender before releasing. The originating sender returns a dial tone to the calling party, and then receives and temporarily stores the digits dialed. As soon as all seven digits have been recorded, the originating sender calls for an idle marker and transmits the called line number to it. In large central offices handling considerable traffic, separate markers are usually provided for dial-tone connections and for completing calls.

Upon receiving the digits of the called number, the marker recognizes from the office code that the call is for another crossbar office. The marker, therefore, seizes an idle outgoing sender, transmits to it all seven digits received from the originating register, and, since an interoffice call is involved, directs the outgoing sender to prepare to transmit only the last four digits corresponding to the called line number. At the same time, the marker selects on the trunk-link frame an idle outgoing trunk circuit to the distant office, and closes a path through the junctor group frame, to connect the calling line on the line-link frame to this selected outgoing trunk on the trunk-link frame. With the completion of these functions, the marker signals the outgoing sender to prepare to pulse out the digits of the called number to the distant central office, and releases.

The incoming trunk in the distant central office is the terminating end of the previously selected outgoing trunk in the calling office. This incoming trunk is connected to the trunk-link frame, and to a register-link frame in the called office (as indicated in Fig. 1-11) which now will be assumed to represent also the called or distant office. When the incoming trunk is seized by the outgoing trunk circuit in the calling office, it selects an incoming register circuit. The incoming register will now signal the outgoing sender in the calling office to pulse out the four digits of the called line number. This information is transmitted in the form of multifrequency (MF) tones and recorded by the incoming register. An idle marker is next selected by the incoming register and the four digits of the called line number and other related data are sent to it. The marker first connects to a number group frame in order to obtain the particular line-link frame and location of the called line number, as previously explained for the No. 1 Crossbar System. Upon receiving this information, the marker connects to the indicated line-link frame and tests the called line number. If it is idle, the marker selects a path through the junctor group frame to connect the called line's location on the line-link frame with the incoming trunk on the trunk-link frame, and then releases. If the called line tests busy, the marker instructs the incoming trunk circuit to cause its associated outgoing trunk circuit in the calling office to send a busy signal to the calling party. The outgoing trunk circuit supervises the call and also controls the disconnection of the operated crosspoint contacts of the crossbar switches on the line-link and trunk-link frames when the calling party hangs up the phone.

The originating sender in the No. 5 Crossbar System is employed for some 15 to 20 seconds on a call, depending on the time needed by the calling customer to dial the called number. The marker requires only a fraction of a second to perform its functions of testing and selecting an idle trunk, connecting links, and actuating the magnets of the selected crossbar switches. A marker is connected to an originating or incoming register by a large number of wires so that the necessary data can be exchanged within a few milliseconds. Thus, a marker can quickly complete a call and be available to handle thousands of calls per hour.

## Electronic Switching Systems

Direct control switching systems, such as the step-by-step type, made up about 52 percent of the Bell System's 16,100 central offices in 1974, and almost all of the approximately 11,200 dial offices operated by non-Bell and Independent Telephone companies in the United States. The 7,800 common control switching systems in the Bell System were divided among 6,350 crossbar, 260 panel, and 1,160 electronic switching types. Electronic switching offices now make up over 15 percent of all common control offices whereas none was in service prior to 1965. It is understood that, in a few years, all panel offices will be replaced by electronic switching types, and that new central office installations will employ electronic switching techniques.

The need for electronic switching derived in part from the slow switching speeds of electromechanical systems, including the crossbar category, and the ever increasing demands by customers for new and improved kinds of telephone services. In electromechanical systems, the development of new and more sophisticated circuits to meet these requirements had been restrained by the inadequate operating speeds of relays and related mechanical apparatus. As a result, research and development activities turned to electronic switching, especially since this technique had been successfully employed in digital computers of various weapons systems during World War II. The invention of the transistor, subsequent improvements in solid-state devices, such as integrated circuits, and the application of computer programming methods to processing telephone calls made possible present electronic switching systems.

The great importance of switching speeds may be illustrated by the fact that several *milliseconds* (one millisecond equals one thousandth of a second) are required in electromechanical systems for the operation of relays to perform switching and related functions. In contrast, transistors, diodes, and other electronic devices can change state or switch circuit conditions at speeds of only a few *nanoseconds* (one nanosecond is equivalent to one billionth of a second or $10^{-9}$ seconds). This ultra-high speed justifies the greater complexity of electronic switching systems. Moreover, these high switching speeds make it feasible for a single common control element, for example, the central control unit in an electronic switching office, to serve as many as 65,000 telephone customers. In contrast, the No. 5 Crossbar System can handle a maximum of only 10,000 lines, and it also requires a considerable amount of duplicate groups of common control equipments.

Other important advantages of electronic switching systems include the anticipated long service life of the equipment, and its capability of adapting to new features and services without changing the system design. This design flexibility is the result of utilizing the stored program and changeable memory concepts for processing calls in electronic switching offices. The stored program feature is based on logic and memory principles employed in electromechanical switching systems. Memory actually is information that is recorded and may be considered as the specific procedure to follow, or ***what*** to do next, in processing a telephone call. Logic makes the decision of ***how*** to do it. For example, memory knows what equipments in the central office should be connected to route a particular call, while logic decides what paths will be used for interconnecting the equipments. As a more pertinent example, in the No. 5 Crossbar System, the originating registers and outgoing senders provide the memory, while the markers make the logical decisions. These precepts will be further explained as we examine the applicable equipments and circuit elements of electronic switching systems in this book.

## Questions

1. What is the designation of the telephone switching network serving North America? How many digits must be sent over this network to reach telephones in Canada and Puerto Rico?
2. What facilities in addition to routing and completing calls are furnished by local central offices? What is the network class designation of an end central office and a toll central office?
3. What is the approximate number of central offices in the United States? What is the theoretical capacity and sequence of line numbers in a central office?
4. What distinguishes an area code from a central office code? What are the area codes for Chicago, Los Angeles, New York, and Washington, D.C.? How many telephone numbering zones is the world divided into?
5. Name three or more important functions performed by the telephone instrument or station subset.
6. How many pushbuttons are provided in present Touch Tone® or touch calling keysets? What particular pushbuttons are used primarily with electronic switching offices? How many audio tones are provided in pushbutton keysets? How many tones are sent for each digit?
7. What are the two general categories of switching systems? What is the main difference between them? Name two or more electromechanical switching systems used now in the larger cities.
8. What is the main difference between the No. 1 and No. 5 Crossbar Systems? What is the principal control device used in both systems? How long is this control device normally utilized for processing a call?
9. What type of switching system is installed in more than half of central offices at the present time? What percentage of central offices are currently using the electronic switching type? What two major disadvantages of electromechanical switching systems instigated the development of electronic switching?
10. What is a nanosecond? Define memory and logic with respect to switching systems and give an example of each.

# 2 CONCEPTS OF ELECTRONIC SWITCHING SYSTEMS

## Design Principles

The development of electronic switching systems has been the result of many new concepts which are entirely different from the designs of electromechanical systems. The relays and crossbar switches of the No. 5 Crossbar System and the selectors and sequence switches of the panel system, for instance, are not replaced by transistors, diodes, or other electronic circuitry. Instead, stored program methods like ones used in computers, including programmed logic and temporary memory techniques, are utilized. The procedure for handling calls, whether originating or terminating, is actually divided into a series of many discrete steps. For example, the processing of an outgoing call starts with the initial handset off-hook condition, and proceeds through various operational steps until the called line is reached and the called party answers the ringing signal. The entire call-handling operation, however, appears to be accomplished almost instantaneously as far as the telephone user is concerned.

The program instructions for handling a call employ the binary symbols and other elements of the machine language (usually designated *software*) that are commonly used with computers. The subsequent translation and interpretation of the program instructions are directed by wired logic in a central control unit of the electronic switching equipment. These operations generally follow the format used in computers and other information processing machines. For instance, information may be shifted from one element or register of the central control unit to another for comparison with information received from a third element. At the same time, the difference between these information inputs or received data may be stored in another register and even read out in still a different one. The central control unit with its wired logic may be considered as the most important concept of common control in electronic switching systems. It governs the basic information processing operations, but not the switching logic which is contained in the stored program as explained later.

## Types of Electronic Switching Systems

Two main types of electronic switching systems for telephone central offices have been evolved and placed in operation during the past several years, the space division and the time division multiplex (TDM). Both employ similar techniques of line and trunk scanning, program instructions, and call store memory although different control and transmission methods are utilized. In the space division method, used in the Bell System's No. 1 and No. 2 Electronic Switching Systems (ESS), the General System's No. 1 Electronic Automatic Exchange (EAX), and other electronic switching systems, an individual path is established between each calling and called line. The TDM system, on the other hand, employs a common path which is time-shared by all connecting lines and trunks through a series of high-speed electronic gates. The Bell System's No. 101 ESS, primarily designed for PBX installations, employs this TDM electronic switching process.

## Principles of Time Division Multiplex Electronic Switching

The common transmission path or highway of time division electronic switching systems may be compared in some respects to time division multiplex (TDM) telephone carrier systems used to provide many voice channels over the same pairs of conductors. For example, in a TDM carrier system, such as the pulse code modulation (PCM) type, a particular voice signal is sampled on a repetitive basis and transmitted in a definite time sequence with respect to the samples of the other voice channels on the same conductors of the cable.

To assure accurate transmission and reproduction of the original speech, the voice signals must be instantaneously sampled at periodic intervals. Also, the sampling rate must be at least twice the highest significant frequency of the voice signal transmitted. In this connection, voice channels normally are designed for a frequency range of about 200 to 3300 Hertz (Hz) or a nominal bandwidth of 4000 Hz (4 kHz). Therefore, a sampling rate of 8000 Hz is usually provided so that two samples will be obtained during each half-cycle of the speech signal. These speech samples are then converted into a series of pulses for transmission over a common path, separate cable pairs being used for each direction of transmission. In a similar manner, a number of telephone calls can time-share the same switching path or transmission highway in a TDM type of electronic switching office.

Figure 2-1 is a simplified block diagram illustrating the time-shared principle used in TDM type electronic switching equipment such as the Bell System's No. 101 ESS. Assuming that an intraoffice connection has been established between calling line A and the called party B, both lines will be periodically connected (but not simultaneously) to a common transmission path through their corresponding two-way electronic gates 1 and 2 in sequence

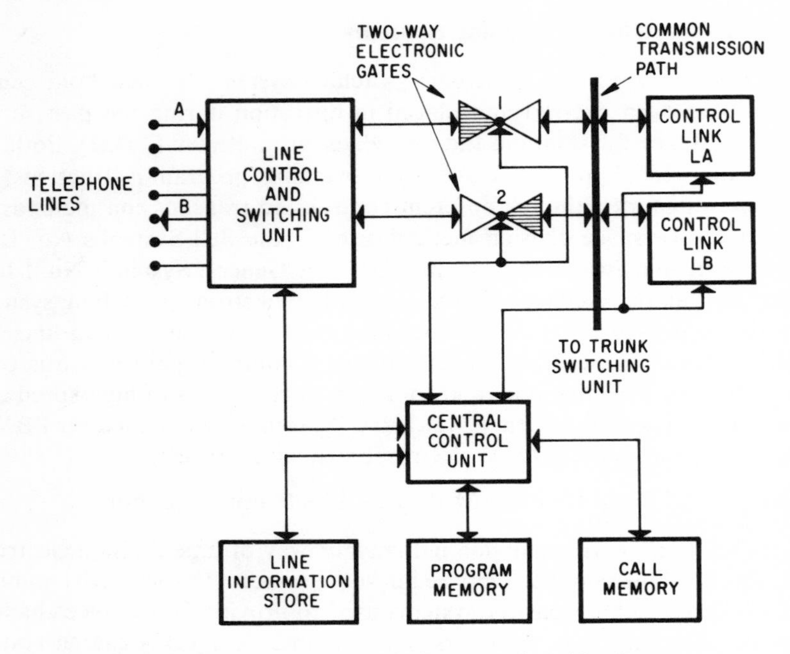


**Fig. 2-1.** Major Elements of TDM Electronic Switching System

with their assigned time slots. Whenever line A is connected in its assigned time slot (by electronic gate 1) to the common transmission path, an idle control link, LA, likewise will be connected to this common path at the same instant under guidance of the central control unit. A sample of the speech signal from line A will be stored temporarily in this link circuit. When called line B is next connected to the common transmission path in its assigned slot by electronic gate 2, control link LA again will be connected. The speech sample previously stored in control link LA will be transmitted to the called line B during its time-slot period. The same procedure will take place for the speech samples of line B for which another control link, LB, would be selected. For calls to or from other central offices, control links associated with the trunk switch units would be utilized in a similar fashion as for intraoffice calls.

The appropriate control pulses to operate the respective two-way electronic gates and the control links are sent by the central control unit, in the proper time-slot intervals. Central control receives the required data for these operations from its associated call and program memory circuits and from the line information store as indicated in Fig. 2-1. In a similar manner, other telephone lines in TDM electronic switching offices can be interconnected utilizing assigned time slots in the common transmission highway. The TDM electronic switching method is satisfactory for PBX installations and small central offices serving a few thousand lines. However, there are many difficulties in adapting the TDM process to the high-calling rate and traffic volume requirements of large central offices. These factors led to the development of the space division method of electronic switching.

## Space Division Electronic Switching Principles

The space division category of electronic switching is exemplified mainly by the Bell System's Nos. 1 and 2 ESS types and General Systems' No. 1 EAX. Although the descriptive data and explanatory information in this book pertain primarily to these types, the principles, techniques, and equipments are applicable in many respects to other space-division electronic switching systems, such as the SP-1 in Canada and the D-10 in Japan.

The principal elements of the space division type, as illustrated by the No. 1 ESS, may be grouped into two main parts: the central processor, and the switching network, with each part functioning separately. The central processor consists of central control, program store, and call store. The following peripheral principal elements are associated with the central processor as shown in Fig. 2-2: line, junctor, and trunk scanners, signal distributors, and the master control center.

There is a continuous exchange of information between central control and its associated call store and program store. This information concerns

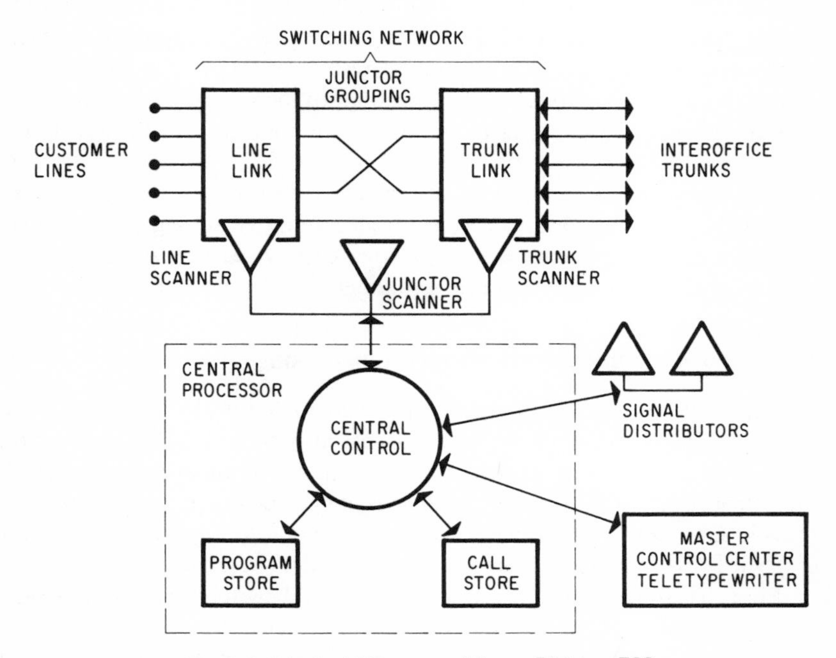


**Fig. 2-2.** Principal Elements of Space Division ESS

calls in progress, particularly the status of customer lines, paths in the switching network, and outgoing trunks. Instructions for guiding the call-processing operation are contained in the semipermanent memory of program store. The call store derives its name from the fact that it temporarily stores information about a call in progress for the guidance of central control.

The switching network, comprised of the line-link and trunk-link units, interconnects calling customers' lines with selected outgoing or intraoffice trunk circuits. The paths selected through the switching network consist of pairs of metallic contacts or crosspoints enclosed in special reedcapsule-type switches called *ferreeds* or *correeds* as explained later. These particular crosspoints, however, do not make or break circuits in which current is flowing. Relays in junctor, trunk, or service circuits provide the final closing or initial opening of the switching path to cause or interrupt current flow.

Ferrod lines, junctor scanners, and trunk scanners are used respectively to constantly monitor customer lines, paths, and trunk circuits for any changes in their status. For example, a customer's line on-hook may indicate a disconnected call, or if off-hook that a call has been initiated. A junctor or trunk circuit may be idle or busy. The conditions observed by the electronic scanners are periodically sent back to the central control unit.

The master control center administers the operation and maintenance of the electronic central office. It includes alarms, lamp displays, control keys, line and trunk test panels, an automatic message accounting (AMA) recording unit, and a memory card writer. A teletypewriter provides information concerning system operations and trouble reports in the form of page printouts. It is also used to type information into the switching system such as temporary translation changes in the call store memory or in operations of central control.

The foregoing presents only a very brief description of the basic elements involved in a small space division electronic switching system. There are many more components and sophisticated devices in a complete operational system such as subsequently explained for the No. 1 ESS.

## Manual and Automatic Switchboard Operations

Explanations of the principal concepts involved in the design and operation of modern electronic switching systems may be simplified by an analogy with respect to the operations of the antiquated manual switchboard. (A similar procedure was followed in discussing electromechanical switching systems in Chapter 1. References also will be made to applicable operations of the No. 5 Crossbar and other electromechanical systems as appropriate.)

First, let us visualize a multiposition manual switchboard controlled by a human operator. Normally, each operator watches the line lamps on the panels as shown in the photograph of Fig. 2-3. Each customer's line is connected to its own jack and line lamp, and to the adjacent four-digit number

**Fig. 2-3.** Manual Switchboard Position (Courtesy of AT&T Co.)

plate which identifies the line number. When a line lamp lights, the operator inserts the plug of an answering cord into the jack below the lamp, obtains the called line number, and completes the call in accordance with the procedure outlined in Chapter 1. Note that an operator devotes complete attention to completing one call before undertaking another one. Other customers initiating calls, therefore, have to wait their turns.

In contrast to a manually operated switchboard would be one operated electronically or automatically. An automaton, because of its almost lightning speed, can handle all positions on the manual switchboard much faster than human beings or mechanical devices. Its electronic eyes (ferrod scanners) periodically scan (about every 100 milliseconds) all line lamps on every switchboard position, one at a time. Whenever a lamp lights, our automaton stops to take appropriate action before proceeding to scan other lamps. For example, assume that line lamp 1234 lights, our electronic operator immediately consults its electronic slate (call store) concerning the previous information recorded about this particular line. Finding that line 1234 previously was idle (on-hook state), our automaton deduces that a call has been initiated, connects dial tone, and continues its scanning operations and other functions.

The digits dialed by the calling customer (for example, 555−4431) are recorded, one at a time as dialed, on the electronic slate by our automaton who also continues to perform scanning and other jobs in the interim. For instance, suppose that line lamps 7890 and 2112 lighted. Our electronic operator would enter this information on other vacant rows on its electronic slate, connect dial tone to these new calling lines, and proceed to record the called line numbers in a similar fashion. These various operations are interlaced with the work of serving other customer lines.

Returning to line 1234, when all digits of the called number (555−4431) have been recorded on the electronic slate, our automaton refers to the information stored in its semipermanent memory (program store) to obtain routing and related instructions concerning the called line number. It then proceeds to complete the call by selecting an idle outgoing trunk in the jack field above the line lamps as shown in Fig. 2-3. At the same time, it erases the information on this call that had been previously recorded on row A on the electronic slate. That row now will be available for use on another call. Figure 2-4 illustrates some of the possible entries that might appear on this fictional electronic slate.

The above analogous representations will help to explain the functions of two essential techniques of electronic switching: the single ultra-speed central control unit represented by an electronic operator or automaton, which is time-shared by all customer lines, and the temporary memory concept (call

| | SQUARE NO. | 1 | 2 | 3 | 4 | 5 | 6 | 7 | 8 | 9 | 10 | 11 | 12 |
|---|---|---|---|---|---|---|---|---|---|---|---|---|---|
| ROW NO. | CALLING LINE NO. | LINE STATUS | CALLED NO. DIALED | | | | | | | | | | * |
| A | 1 2 3 4 | 1 | 5 | 5 | 5 | 4 | 4 | 3 | 1 | | | | 1 |
| B | 4 5 7 6 | 0 | | | | | | | | | | | |
| C | | | | | | | | | | | | | |
| D | 7 8 9 0 | 1 | 2 | 0 | 1 | 5 | 5 | 5 | 1 | 2 | 1 | 2 | 0 |
| E | | | | | | | | | | | | | |
| F | 2 1 1 2 | 1 | 6 | 3 | 7 | 1 | 2 | | | | | | |
| G | 4 5 7 6 | 1 | 3 | 1 | 2 | | | | | | | | |
| H | | | | | | | | | | | | | |
| I | | | | | | | | | | | | | |
| J | 4 4 5 8 | 1 | 0 | | | | | | | | | | 1 |
| K | | | | | | | | | | | | | |

LINE STATUS:
0 = ON-HOOK
1 = OFF-HOOK

*DIALING STATUS:
0 = AREA CODE DIALED
1 = DIALING COMPLETED

**Fig. 2-4.** Analogous Electronic Slate

store) illustrated by the erasable electronic slate. The same manual switchboard analogy can be further utilized for explaining the functions of the semipermanent memory or program store. For instance, to handle the many diverse types of telephone calls, it would be necessary for our electronic operator to memorize what precise steps to take in processing the various classes of calls. Any changes in the established procedure, however, would necessitate retraining our automaton. A more effective and simple method is to provide a step-by-step program of instructions or an operating manual. This manual could contain simple instructions for handling calls such as, "Scan line lamp 1234. If it is lighted, write 0 in row A, square 1 on the electronic slate. If it is not lighted, write 1 in the same place on the electronic slate." Or, "Test line 7890 for busy or idle condition. If busy, refer to page 73, paragraph 37.1. If idle, refer to page 38, paragraph 16.4." In this manner, our electronic operator has to be trained only to execute the indicated steps of instructions written in the manual (program store). To change the procedure for any type of call, it is necessary only to write a new manual or program using the same type of instructions. Our electronic operator will continue to function in the prescribed manner since the knowledge for processing calls is stored in the instruction manual (program store) and not in its memory.

In the foregoing analogy to the manual switchboard, we have assumed an electronic operator (central control) using one electronic slate (call store) and an instruction manual (program store). Ideally, a single automaton is capable of working continuously 24 hours each day. To ensure continuity of operations, however, another automaton with its own electronic slate and instruction book is provided. Both automatons work together on every call, but one is placed in charge to actually control calls such as manipulating the switchboard cords and keys. The two automatons watch the same line lamps and other lamp signals, both independently make entries in their electronic slates and refer to their instruction manuals. In this way, cross-checks are made continuously between the functions accomplished by our two electronic operators. Thus, any differences or possible trouble indications in the instruction manual, electronic slate, or an automaton's operations will be immediately disclosed.

In the event of a malfunction, our electronic operators would follow the related instructions in the manual. If one electronic slate should become defective, both automatons can read from the other one although only the automaton in charge can make entries in the working electronic slate. Similarly, if the instruction manual needs repair, both electronic operators can read from the other one. Likewise, it is possible for one electronic operator to utilize both electronic slates and instruction manuals. Moreover, any trouble conditions encountered would be immediately referred by our automatons to the local test desk (master control center) in accordance with the instruction manual. Thus, maintenance personnel will be kept informed of the situation and will be able to initiate the required remedial measures.

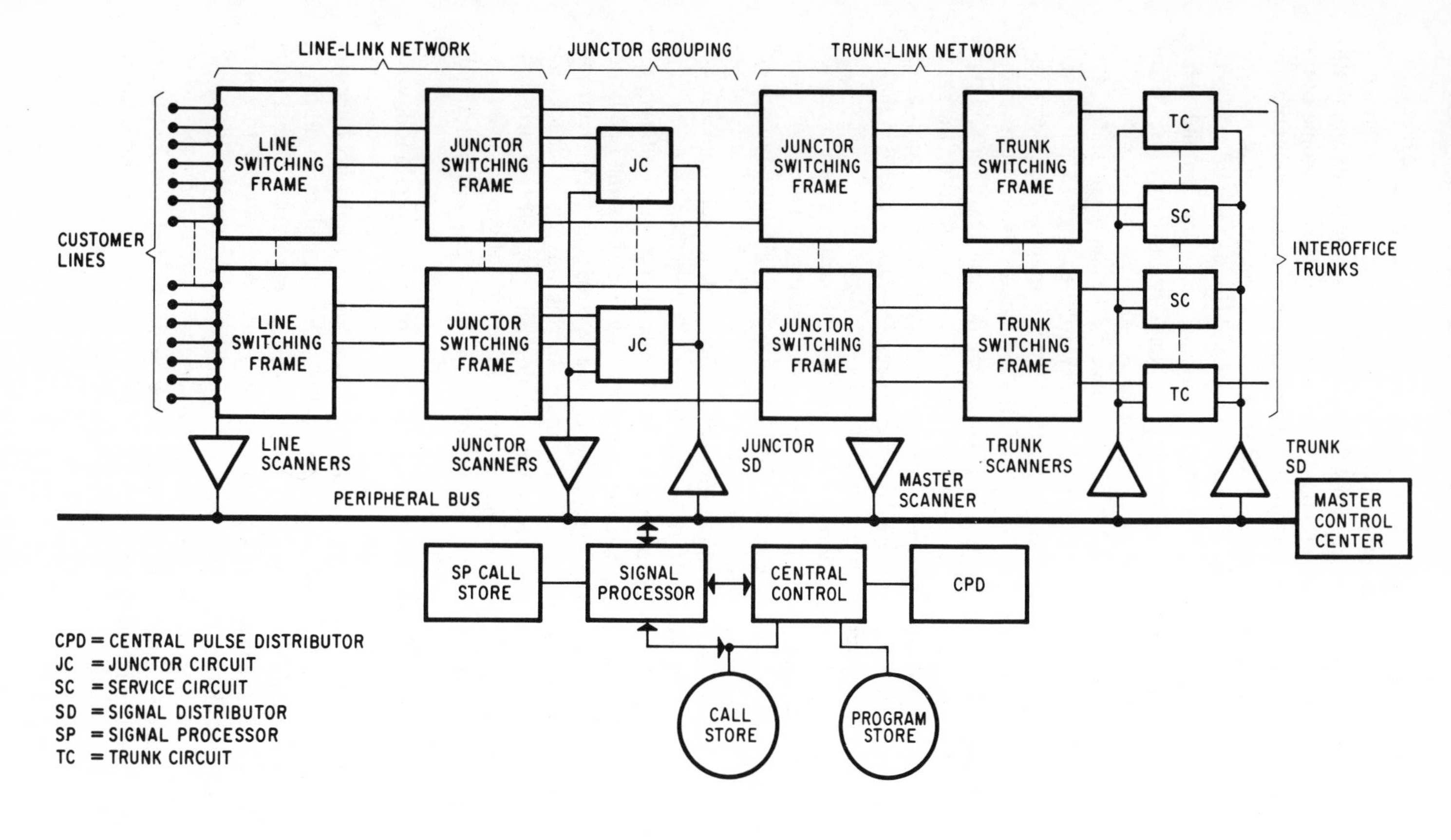


**Fig. 2-5.** Basic Elements of ESS

## Functions of Basic Electronic Switching Elements

Various sophisticated components and special devices have been developed for electronic switching systems. Figure 2-5 is a block diagram of such elements used in the No. 1 ESS. Brief descriptions of these elements and their functions follow.

### *Line-Link and Trunk-Link Networks*

The switching network in the No. 1 ESS is formed by the line-link and trunk-link networks with the interconnecting junctors on the junctor grouping frame. This arrangement, which corresponds to the jacks and cord equipment on the manual switchboard in the analogy of our electronic operator, is shown in Fig. 2-6. Each switch frame has a two-stage network so that the complete switching network comprises eight stages. The actual switching operations are performed by unique crosspoint ferreeds which are a form of miniature glass-enclosed reed switches. These reed switches are operated or released by controlling the magnetization of two adjacent metallic plates made of Remendur, as explained later. The switching network has a maximum capacity of 65,000 lines and 16,000 trunks.

A line-link network contains four line switch frames and four junctor switch frames. Each line switch frame includes the ferrod sensors and bipolar ferreed switches associated with the customer lines that it serves. Normally, sixty-four customer lines have access to sixteen links for a 4 to 1 concentration ratio. The particular concentration ratio employed depends upon the

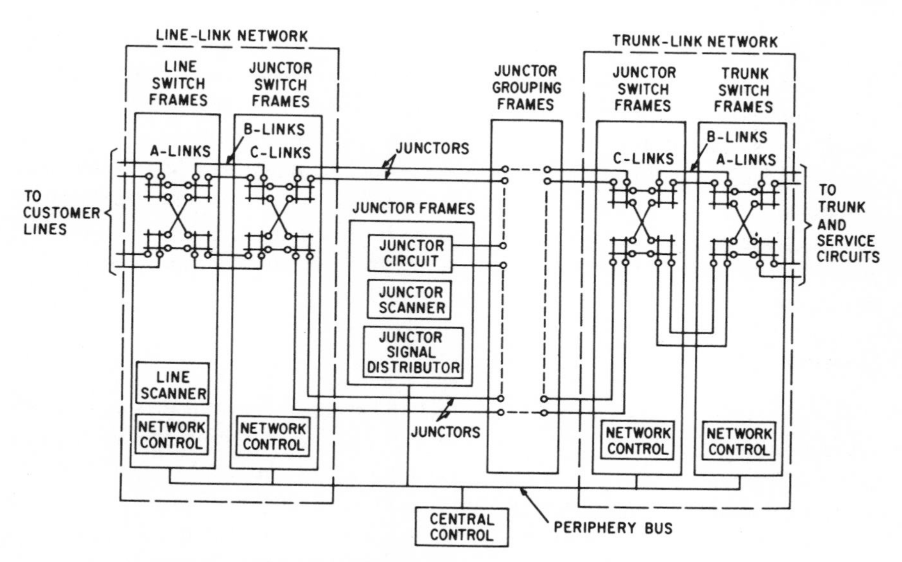


**Fig. 2-6.** Switching Network in Electronic Central Office

type of customer line. For example, business and PBX lines, because of their higher calling rates, may be arranged for a 2 to 1 concentration ratio while the service needs of residential customers may be met with a 4 to 1 ratio.

A trunk-link network usually has four junctor switch and four trunk switch frames. The number of these frames can be increased if the traffic volume requires a larger concentration ratio. For the usual 1 to 1 concentration ratio, one trunk-link network will have a capacity of 1,024 trunks, the same as the number of junctors. For a 2 to 1 ratio, the number of trunks can be increased to 2,048.

The crosspoint ferreeds on the various switch frames perform all path interconnections required for a call under direction of central control. Note that only two-wire paths, the tip (T) and ring (R) conductors which comprise the transmission circuit, are switched by the actions of the ferreeds.

### *Junctor and Junctor Grouping Frames*

The line-link network connects through wire junctors (pairs of tip and ring conductors) to the trunk-line network. The wire junctors run through the junctor grouping frames, as shown in Fig. 2-6. In addition, junctor circuits on the junctor frame would be connected to the line-link network to provide transmission and supervision on intraoffice calls. Junctor frames also are equipped with junctor scanner and signal distributor circuits. Figure 2-7 shows a typical junctor frame with its jack and plug interconnections.

### *Switching Network*

The switching network with its many links and paths available for handling calls, requires some means to denote which paths and trunks are idle, and which are busy at any given time. In electromechanical switching systems, use is made of the S (sleeve) or C (control) leads as a memory device for this purpose. Selectors in the step-by-step and panel offices test these leads when hunting for idle paths. In the No. 5 Crossbar System, the *marker* performs this function. Electronic central offices make use of memory in the call store to select idle links, paths, and trunks in a special way, as subsequently explained.

### *Scanner Circuits*

Scanner circuits are used to detect the state of lines and trunks. Scanners are associated with the line-link, and trunk-link switching operations and junctor frames. Scanners are also connected with trunk and service circuits. A master scanner controls the line and trunk scanning circuits.

The scanning device used is a ferrod sensor. Its design and operation are explained later. Line scanners are utilized to periodically check the condi-

Fig. 2-7. Jack and Plug Interconnections on Junctor Frame (Courtesy of Bell Telephone Laboratories)

tions of customer lines to determine, for example, whether an off-hook or on-hook state exists. These scanner operations are directed by central control in accordance with instructions received from program store in connection with call-processing. Junctor scanners observe the condition of junctor circuits, and trunk scanners monitor the trunk and service circuits. These various scanners may be compared, referring to the manual switchboard analogy, to the eyes of our operator.

Scanners perform their different functions in periodic intervals governed by central control. The scanpoints, or points to be scanned, are specified in the information (*address signals*) received from central control. The conditions observed at the specified scanpoints by the ferrods of the scanners are sent back to central control.

A photograph of a ferrod used as a line scanner is shown in Fig. 2-8, and its schematic diagram is outlined in Fig. 2-9. The ferrite rod has two control windings, an interrogate winding used to determine the state of the circuit, and the readout winding which advises central control of the circuit's

**Fig. 2-8.** Ferrod Line Scanner (Courtesy of Bell Telephone Laboratories)

condition. These four windings over the ferrite rod comprise a transformer. Current flowing in the two control windings determines the strength of the magnetic coupling between the interrogate and readout windings, each of which is a single turn of wire threaded through two holes in the center of the ferrite rod. When a pulse is applied to the interrogate winding while current is flowing through both control windings (off-hook or closed loop condition) no signal will be induced in the readout winding because the ferrite rod will be saturated by the resultant magnetic flux. The no-signal readout, therefore, is termed "0." If current is not flowing through the control windings (handset on-hook or open loop condition) when a pulse is applied to the interrogate winding, the ferrite rod will not be saturated. In this case, a pulse will be induced in the readout winding, and the readout is said to be "1." Consequently, whenever an interrogate pulse is applied to a ferrod, the occurrence

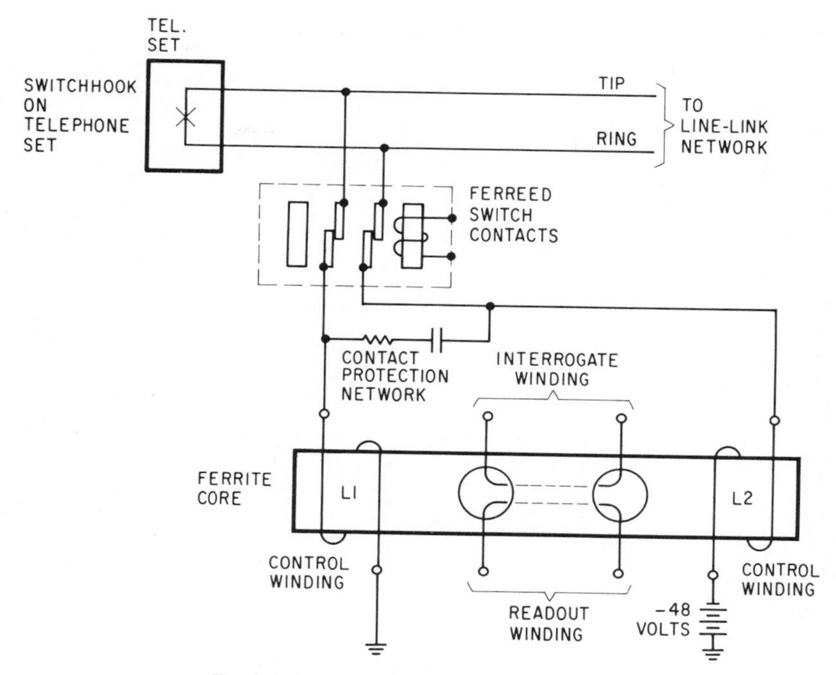


**Fig. 2-9.** Ferrod Used as Line Scanner Circuit

or absence of a readout pulse denotes whether the circuit being scanned is open (readout = 1) or closed (readout = 0).

## *Signal Distributors*

Signal distributors furnished in junctor and trunk frames are used to operate or release magnetic latching relays in these circuits as directed by central control. In fact, signal distributors serve as buffers between the high-speed operations of central control and the low-speed electromagnetic relays. With respect to the aforementioned analogous manual switchboard, signal distributors may be considered as the hands of our operator handling the cords and keys on switchboard positions.

A signal distributor is divided into two parts, and each part comprises a controller which receives and stores information or address signals from central control. The two controllers operate independently but if one should fail, the other can take over control of the same signal distributor. The controller connects to the specified latching relay's winding and applies the required current pulse to operate or release the relay. Approximately 20 milliseconds are required by the controller to complete its functions. A latching relay has a

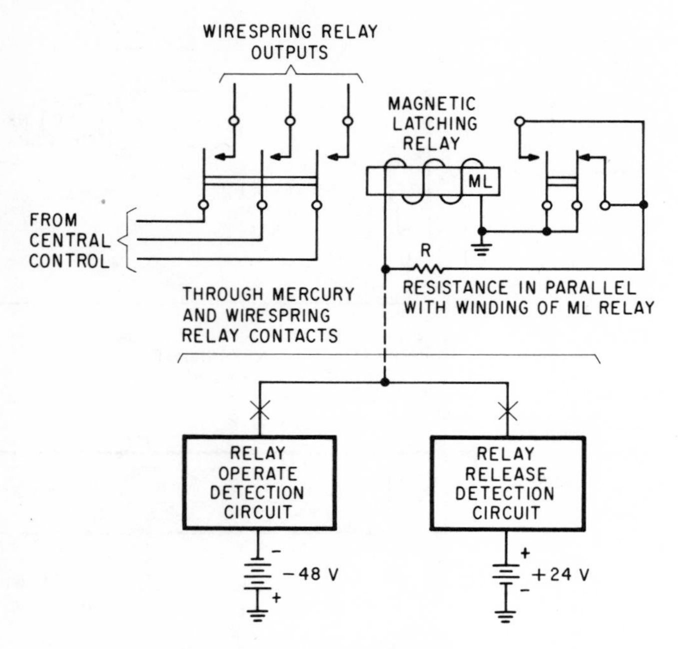


**Fig. 2-10.** Magnetic Latching Relay Circuit

remanent core material which can retain sufficient magnetism to hold the relay operated after the operating pulse (−48 volts) is received. The latching relay releases on a current pulse of opposite polarity (+24 volts). Thus, only one conductor is needed to operate or release each magnetic latching relay.

Figure 2-10 is a simplified schematic drawing of a magnetic latching relay circuit. The relay operates on a −48 volt pulse and releases on receipt of a +24 volt pulse. The resistor, R, which is in parallel with the relay's winding, is momentarily disconnected by the contacts of the relay whenever it operates or releases. This results in a change in the current flowing through the latching relay's winding. The associated controller circuit in the signal distributor detects this current change and informs central control.

## *Central Pulse Distributors*

Central pulse distributors are used to transmit high-speed pulses to control the peripheral units such as scanners, signal distributors, relays in trunk and service circuits, teletypewriters, and automatic message accounting recorders in the master control center. The functions of a central pulse distributor likewise may be compared to the hands of our operator in the operation of the analogous manual switchboard. Central pulse distributors are provided in pairs, up to eight, and the units in each pair operate independently of each other.

Information (address signals) is sent by central control to a central pulse distributor which, in turn, transmits appropriate pulses to many points in the electronic switching system over a common group of wires called the *peripheral bus*. The particular peripheral unit, such as a signal distributor, that is required to respond to this information on the common peripheral bus is selected by an enable pulse from the central pulse distributor. The time interval between consecutive orders from central control to a central pulse distributor is a minimum of 11 microseconds.

## *Trunk, Junctor, and Service Circuits*

Trunk, junctor, and service circuits are associated with the switching network. They are utilized to complete and to supervise the paths that have been established through the switching network by central control. These circuits normally operate under the direction of central control through a central pulse distributor or signal distributor. Central control detects, by means of associated ferrods or scanners, any change in the trunk, junctor, or service circuits caused by relay operations, on-hook or off-hook actions by the customer, or a change of circuit or loop conditions made by the distant central office.

Trunk circuits only provide supervision and transmission functions. They are classified as outgoing, incoming, or two-way trunks depending on whether the local, distant office or both can connect to the particular trunk circuit. Connections between trunk and service circuits are made through the trunk-link network. The components in a trunk circuit may be reduced to only a pair of wires to permit an associated service circuit to handle pulsing and supervisory signals. In this case, called the bypass state, the transmission elements, such as the repeat coil, are switched out, thereby providing a metallic path through the switching network.

Figure 2-11 is a simplified schematic drawing of an outgoing trunk circuit that is connected through the switching network to a customer's line. The trunk scanner normally scans ferrod sensors at scanpoints A and B in the circuit. If the calling customer's line is off-hook, current will flow through the two control windings of the ferrod at scanpoint A. The trunk scanner, therefore, will readout "0" for the same reasons as were discussed previously under scanner circuits. If the calling line is on-hook, no current will flow through the control windings and the trunk scanner will readout "1."

A supervisory battery is connected to the trunk conductors by the distant central office and this voltage will be applied to diodes D1 and D2 in the outgoing trunk circuit. If the trunk loop is normal (battery on the tip conductor and ground on the ring side) diode D1 will have a low resistance and current will flow through the control windings of the ferrod at scanpoint B. A high resistance will be presented by diode D2 under this battery polarity. The trunk scanner, therefore, will readout "0" at this time. When the called party

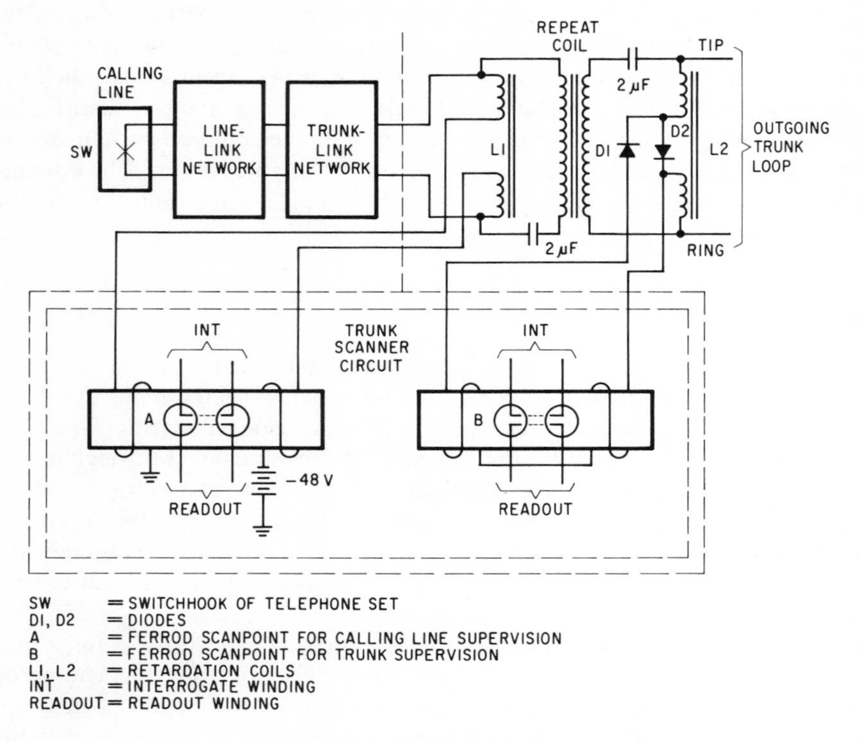


**Fig. 2-11.** Outgoing Trunk Circuit

answers, the equipment in the distant central office will reverse the battery polarity on the trunk conductors. Current will no longer flow through the control windings of the ferrod at scanpoint B because D1 now will have a high resistance and diode D2 a low resistance. As a result, the trunk scanner at scanpoint B will readout 1. During the conversation period, the ferrods at scanpoints A and B will be saturated by the respective loop currents and both readouts will be 0. When either the calling or called party hangs up (on-hook state), central control is informed of this change-of-state occurrence at scanpoints A and B by the trunk scanner, and disconnection will take place.

Junctor circuits are employed on intraoffice calls or connections between calling and called customer lines in the same electronic central office. The junctor circuit, which supplies a talking battery to each line, also contains two magnetic latching cut-through relays, CT-1 and CT-2, as shown in the simplified schematic diagram of Fig. 2-12. These relays, which are controlled by the signal distributor, are used to protect the ferreed switches in the switching network by connecting the battery potential after the switching network path has been closed, and removing the battery circuit before the network path is opened on disconnection.

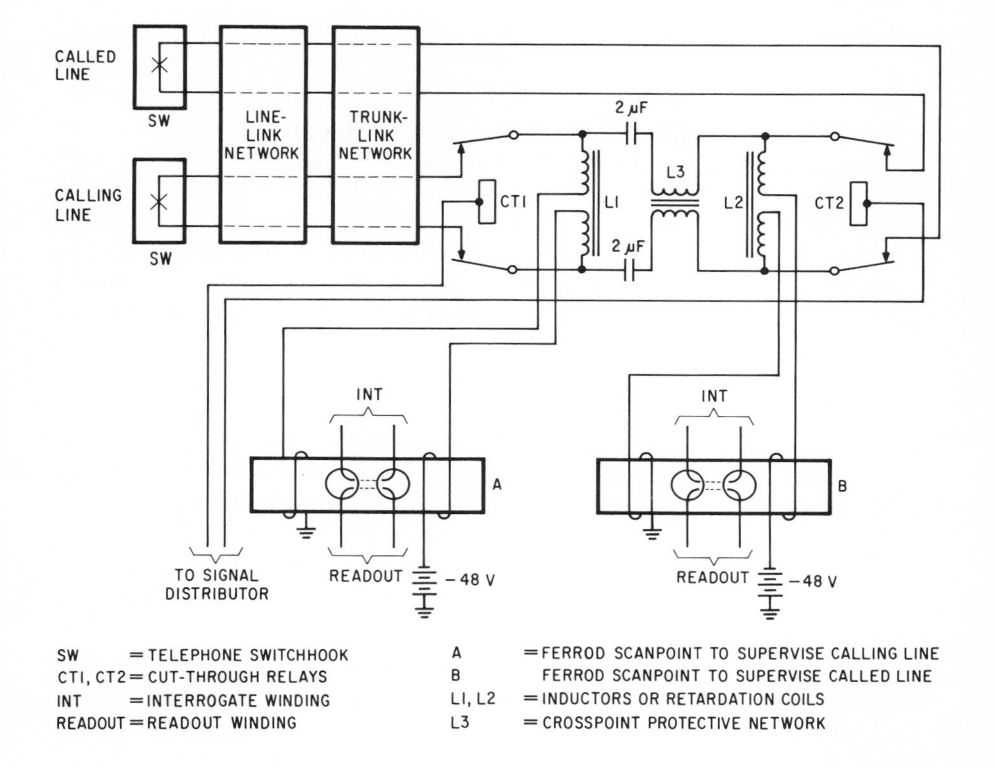


**Fig. 2-12.** Junctor Circuit for Intraoffice Calls

Service circuits are associated with the trunk-link network for connection to customer lines or to trunks as required. They may be considered as accessory circuits to perform functions which can be handled more efficiently by a few special circuits than by providing additional equipment in each trunk circuit. Some examples of service circuits are: the customer dial pulse receiver circuit which supplies the dial tone to the calling line and detects the dialed digits; ringing circuits to ring the called line, and to disconnect the ringing current when the called party answers; and multifrequency (MF) receivers to connect to an incoming trunk circuit in order to receive MF signals from the distant central office, and to convert these tone signals into a form that the particular scanner can observe in order to inform central control.

## *Central Control*

The intelligence for controlling the operations of an electronic switching central office is in the central processor, a high-speed computer and data-processing unit. It may be compared to the brain of our operator in the manual switchboard analogy. The three basic components of the central processor are central control, program store, and call store as illustrated in Fig. 2-2.

Central control is a computer-type binary digital mechanism that conducts complex logic operations in accordance with instructions in its semi-permanent program store and temporary call store memories. The processing logic, such as what paths to interconnect, is contained in the many thousands of logic circuits in central control. For instance, central control decodes and executes binary encoded instructions received from its associated program store at the rate of one instruction every 5.5 *microseconds* (one millionth of a second). This rate is established by clock pulses originating in a 2 MegaHertz crystal oscillator circuit.

Three major classes of instructions normally are processed by central control:

1. The status of lines and trunks. For example, central control may be requested to examine the ferrod line scanners associated with customer lines for any change in condition, such as off-hook or on-hook.
2. Manipulating data received from the program store unit and temporarily depositing the results in call store for recall whenever needed.
3. Generating outputs for operating latching relays in junctor, trunk, and service circuits.

In processing instructions, central control requests applicable information from its program and call stores. For example, upon receiving a request from call store for dial tone connection to a calling line, central control would ask program store for the processing instructions provided in a particular location or address in program store. The instructions sent back by program store would advise central control what to do with the information being processed and, at the same time, indicating the address or where to do it. Figure 2-13 is a photograph of central control (first four frames left to right) followed by central pulse distributor, master scanner, and call stores (last two frames).

## *Signal Processor*

The signal processor with its associated call store is furnished only in the Bell System's larger size No. 1 ESS central offices. It operates simultaneously with and independently of central control. The prime purpose of the signal processor is to relieve central control of many time-consuming repetitive tasks. Thus it serves to increase substantially the traffic-handling capacity of the No. 1 ESS central office. Depending on the size of the central office, up to two pairs of duplicate signal processors may be installed.

Just as with central control, the signal processor is a stored program-type of data processing machine utilizing computer technology. It operates under the guidance of program instructions stored in its call store unit, and

**Fig. 2-13.** Equipment Frames of Central Control, Central Pulse Distributor, Master Scanner, and Call Stores (Courtesy of Bell Telephone Laboratories)

performs both supervisory and nonsupervisory tasks. The supervisory functions comprise the periodic scanning of line ferrods to detect originating calls, junctor circuits to detect disconnection of calls, and trunk circuits for connection and disconnection signals.

The nonsupervisory tasks of the signal processor include the reception of signals from and the transmission of signals to other central offices, receiving Touch Tone® signals or dial pulses from calling customers' lines, detection of permanent signals on customers' lines, and the execution of program orders from call store. The signal processor accumulates in its call store information that is to be further processed by central control. Moreover, central control can transfer information to the signal processor's call store whenever the services of the signal processor are required. The temporary memory in the signal processor is similar to that of central control except that the signal processor employs its temporary memory both as a program store and a call store function.

## *Program Store*

The program store is a semipermanent memory device that contains specific instructions governing the step-by-step operations of central control. It may be likened to the instruction manual used by the operator in our switchboard analogy. Figure 2-14 is a photograph of the program store modules in a typical electronic switching office.

The data contained in program store includes information for translating a customer's line number into an equipment location on the line-link frame and vice versa, equipment locations of trunks on the trunk-link frame, class of service of various customer lines, routing and charging information, and other items concerning particular lines or trunks. The information is contained in a random-access semipermanent memory employing twistor wires and tiny magnets on aluminum cards or modules.

Each aluminum card has sixty-four columns or forty-five tiny spots of magnetic material called *vicalloy*. A column represents the location of a bina-

**Fig. 2-14.** Program Store Equipment Frames (Courtesy of Bell Telephone Laboratories)

**Fig. 2-15.** Twistor Aluminum Memory Card Used in Program Store (Courtesy of Bell Telephone Laboratories)

ry "word" (a series of 0 and 1 characters) which is used to represent stored information in the binary number system as later explained. The forty-five magnetic spots in each column, of which forty-four spots are utilized, correspond to binary digits or bits; a magnetized spot represents a 0 and a demagnetized spot denotes a 1. A photograph of a twistor aluminum memory card used in program store is shown in Fig. 2-15.

The memory contents of the program store can be read out only. The insertion of information, termed writing, into program store necessitates the magnetizing or demagnetizing of the individual tiny magnets in each column of an aluminum card by an external card-writer unit. Electrical malfunctions in circuits will not alter the information contents of the program store.

## *Call Store*

The call store is an erasable or temporary memory unit that is primarily used to temporarily record various types of information pertaining to a call in progress. It corresponds, in our switchboard analogy, to the electronic slate used by the electronic operator.

Data temporarily deposited in call store may include the digits dialed by a calling customer, idle or busy status of links in the switching network, results of maintenance or diagnostic tests, charge information to be recorded on the automatic message accounting (AMA) magnetic tapes, and the digits to be transmitted on an outgoing call. In addition, changes in translation information, such as routing calls to a particular central office, can be temporarily stored in call store until these instructions can be placed into the semi-permanent memory of program store.

The basic memory device employed in call store consists of ferrite sheets, each about an inch square and containing a 16 × 16 array of tiny holes. Every hole has the equivalent of a copper wire threaded through it thereby providing a square-loop memory core as will be explained. The information temporarily stored in call store is arranged in 8,192 words of 24 bits each, which can provide a total storage capacity of 196,608 binary digits (bits). The location of every word in call store is singularly identified by an address as subsequently explained.

Instructions or inputs received from central control specify whether interrogating (reading) or inserting information (writing) operations are to be performed, and the address of the particular location involved in call store. In the case of writing operations, the particular information or word to be stored (written) also is transmitted by central control.

### Master Control Center

The master control center is the operational, maintenance, and administration core of the electronic switching central office. It serves as the interface between the electronic switching equipment and the maintenance personnel. The master control center has facilities for testing customer lines and trunks, alarms to indicate malfunctions, the means for system testing and control of operations, magnetic tapes for recording automatic message accounting (AMA) data for billing purposes, and special equipment for writing information on the memory cards used in program store.

Five separate parts comprise the master control center. The following is a brief summary of their functions:

1. *Alarm, display, and control circuits* that provide continuous indications of system operations, and permit maintenance personnel to control operations of the central office.
2. *Trunk and line test facilities* for maintenance of trunks, lines, and service circuits, including transmission tests, handling permanent signals on customer lines, make-busy circuits.
3. *Teletypewriter channels* for communications between the electronic switching equipment and maintenance personnel, and for traffic measurements, line assignments, etc.
4. *AMA recording charges* on customer calls by providing magnetic tape records of all data pertaining to such calls. The magnetic tape is subsequently processed in an electronic data processing center.
5. *Program store card writer* for periodically updating translation information in the program store.

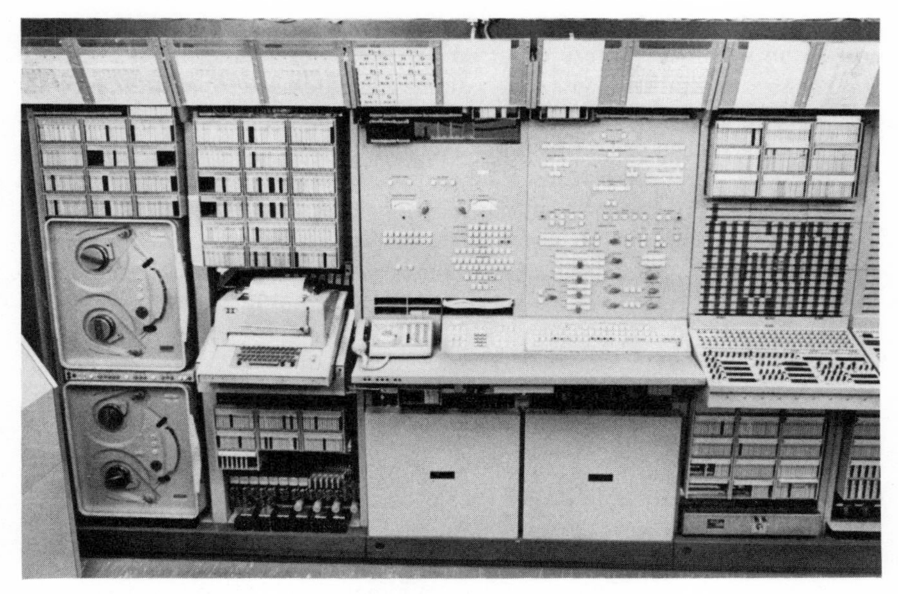

**Fig. 2-16.** Maintenance and Master Control Center (Courtesy of Bell Telephone Laboratories)

Figure 2-16 shows a maintenance or master control center with the aforementioned equipment items. The automatic message accounting (AMA) magnetic tapes are in the first frame on the left, the teletypewriter and related apparatus are in the second frame, the alarms, control, trunk, and line testing facilities are on the other three frames.

## System Operational Reliability

Electronic components, switching equipment, and related circuitry in electronic switching systems are designed for continuous 24-hour operation. Thus, many years of service normally can be anticipated without interruptions because of equipment malfunctions or component failures. To ensure utmost reliability of operations and to provide for necessary maintenance and diagnostic testing requirements, the major common control elements are fully or partially duplicated.

The fully duplicated units are central control, program store, call store, central pulse distributor, and, when equipped, the signal processor with its own call store. Partially duplicated elements are the line, junctor, and trunk scanners, the junctor and trunk signal distributors, and the control circuits of the switching network.

Two central control units actually process the data on every call simultaneously and match circuits are used to compare the results at key points in the system. In case of an error or fault, central control usually repeats the operation on which the failure occurred. A continuing fault will cause central control to attempt an alternative routine or processing operation. For instance, if an error appears to originate from program store, central control may utilize a special program stored in call store which is designed to disclose malfunctions in program store. If these procedures are unsuccessful, central control will try other means, such as utilizing different combinations of duplicated units and peripheral buses, to make the system operational. Related alarm signals are generated at the maintenance or master control center to immediately alert maintenance personnel of these malfunctions.

## Customer Special Services

Another outstanding feature of electronic switching systems is the capability of providing special optional services for customers by means of instructions stored in the memory cards of program store. The following are brief descriptions of some of the special services that are available in electronic switching offices. To provide similar optional features (other than PBX hunting) in electromechanical central offices would require installing special apparatus including electronic equipment, and making extensive wiring changes in the existing common control and switching equipments of these central offices.

### *PBX Hunting*

In S × S and panel type electromechanical offices, PBX lines (Private Branch Exchange installed on a customer's premises and served by several central office lines) must have a consecutive sequence of directory numbers, for example, 5400, 5401, 5402, 5403, 5404, and 5405. This numbering sequence is necessary in order that the S × S connector switch or the final selector in the panel office can hunt for an idle line in the particular PBX line group, in case the first or next several lines are busy. A call to the first line in the group, 5400, can hunt through the entire group if necessary but a call to 5402, for instance, can hunt only to higher numbers and cannot hunt back to lower ones. In the No. 5 Crossbar System, it is not necessary that all PBX lines have consecutive directory numbers as long as they are in the same number group frame. A total of 1,000 consecutive directory numbers, 0000–0999, 1000–1999, 2000–2999, etc. are served by one number group frame.

The above inherent limitations of PBX hunting are not imposed on electronic switching systems. The number of lines in a PBX group and the order of hunting are stored in the memory of program store, and are not dependent on circuit wiring. An incoming call can be made to hunt in any desired sequence through one or more groups of PBX lines. If all lines are busy in the first PBX group, the call can be transferred to another group of PBX lines for subsequent hunting.

## *Class of Service*

Increased versatility is also provided by the stored memory in electronic switching offices for the various classes of service and features available to customers as compared to electromechanical central offices. Approximately 100 classes of service are provided by the wired equipment in the No. 5 Crossbar System, while a total of 1,024 different service classifications can be stored in the program store memory. This large number of service classes permits the assignment of various optional services to customers as may be requested. Each optional or special feature can be assigned a specific class of service indication which could be allowed on some customer lines and denied to others.

For instance, certain PBX lines can be restricted to making calls only within the PBX, others to only local calls, some to toll calls within a particular area or to specified central offices within an area, and certain other PBX lines may not be limited in any manner. These operational functions are directed by central control when it is informed by the translation of the customer's line equipment number in program store concerning what special services the customer is entitled to.

## *Abbreviated Dialing*

Abbreviated dialing is a special service furnished by electronic switching offices. It enables a customer to reach frequently called numbers by dialing only a two or three digit code instead of the usual seven, ten, or more digits. The abbreviated code is not a contraction of the called number but comprises the two-digit prefix "11" (digit * when pushbutton dialing is furnished) and another digit. The 11 or * prefix informs central control that the customer desires to use the abbreviated dial service and that only one or two more digits should be expected. A list of telephone numbers, up to a maximum of 32, which the customer may call by abbreviated dialing, is stored in the memory of program store. Customers desiring this service are assigned a one-digit number for up to ten frequently called numbers, and a two-digit number for eleven to thirty-two frequently called numbers.

Figure 2-17 depicts the steps involved in the abbreviated dialing operations. In this example, it is assumed that the calling customer dials the abbreviated dialing code *7 (corresponding to 117) which is temporarily deposited in call store by central control. Since 117 is recognized by central control as an abbreviated dialing code, this code and the calling customer's line number 7890 and its equipment location numbers are referred to program store for necessary translation. Program store finds that line number 7890 has a list of nine frequently called numbers and that code 117 corresponds to directory number 312−555−1212, and advises central control accordingly. Central control transfers this number (312−555−1212) to call store where it replaces the original dialed code 117. Central control then proceeds to process and

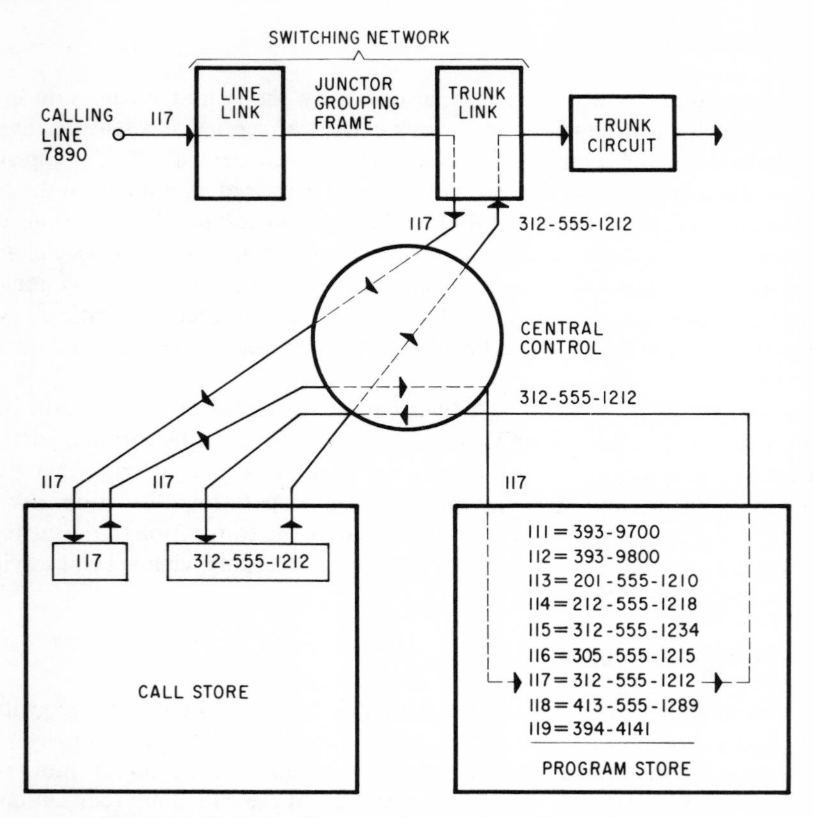


**Fig. 2-17.** Simplified Diagram of Abbreviated Dialing Operations

complete the call to number 312–555–1212 just as if that number had been initially dialed by customer 7890.

## *Automatic Call Transfer*

A customer who will be away from his home or business telephone for the day or evening can have all incoming calls automatically rerouted to another telephone number in the local area. To initiate this service, the customer first dials the assigned two- or three-digit activate code for this particular service. Upon receiving a dial tone again, he dials the directory number of the distant telephone line to which calls are to be transferred. A confirming tone will be heard as an indication that the call transfer service is activated and that all incoming calls will be rerouted to the other telephone number. The call rerouting action is directed by central control in a manner similar to that explained for the abbreviated dialing service and illustrated in Fig. 2-17.

It still will be possible, during the period of this incoming call transfer service, to originate calls from the customer's home telephone whether or not the distant telephone is in use. When the customer returns home or to the office, he can discontinue the automatic call transfer service by dialing the assigned deactivate code and receiving the confirmation tone. Subsequent incoming calls now will connect only to his own telephone line.

A preset automatic transfer of incoming calls also may be utilized. This optional service permits a customer to have all incoming calls automatically transferred to any one of eight other previously selected telephone numbers. These selected numbers are inserted in the memory of program store for the particular customer desiring this service. Consequently, any additions, changes, or deletions in the list of selected numbers must be recorded in program store by the card-writer equipment.

To inaugurate this service, the customer first dials the aforementioned activate code and when the dial tone again is heard, dials one digit to designate to which of eight telephone numbers all incoming calls should be transferred. A confirmation tone is sent to the customer as an indication that the preset transfer service is in effect. The transfer of incoming calls is effected by central control in a similar manner as previously explained for the abbreviated dialing case. To discontinue this service, the customer dials the assigned deactivate code, hears the confirmation tone, and then hangs up.

### *Add-On Conference Call*

Another important service provided by electronic switching offices is the add-on conference call. A customer in conversation with another party can add on a third party to make a three-way conference call. For example, assume that customers A and B are conversing and that A desires to add customer C to the connection. Customer A first flashes the switchhook (equivalent to dialing 1). This momentarily releases the established connection but customer B is still held connected by his associated trunk circuit. Central control recognizes the momentary on-hook state or dial 1 digit as a request for

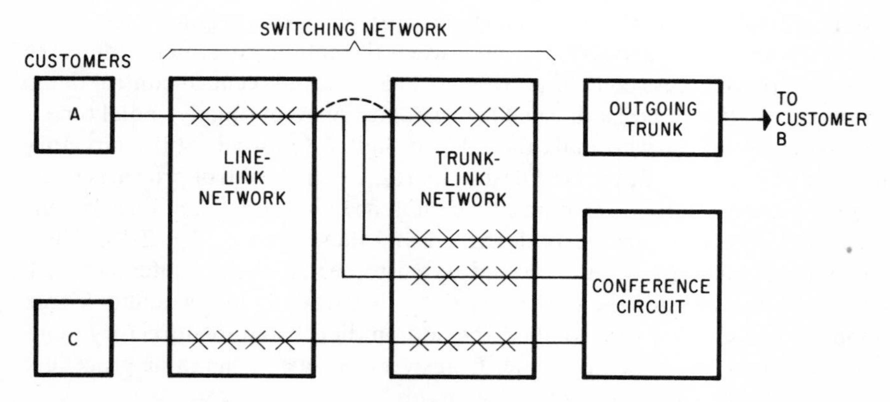


**Fig. 2-18.** Add-on Conference Connections to Switching Network

special service, and connects a digit receiver and dial tone to the line of customer A.

Customer A next dials the add-on digit 2 followed by the directory number or (if he has abbreviated dialing service) the abbreviated code of customer C. Central control then proceeds to connect the switching network paths of lines A, B, and C to an idle conference circuit. Both A and B will hear audible ring or a busy tone depending on the state of line C. When customer C answers, all three parties will converse through the facilities of the conference circuit as shown in the block diagram of Fig. 2-18. The dotted line between the line-link and trunk-link networks in this figure indicates the original talking connection between customers A and B before connection to the conference circuit.

The add-on party, such as customer C, can be removed from the connection at any time by customer A. For this purpose, customer A first flashes his switchhook (dials 1) as in the initiation of the add-on conference service, and upon receiving dial tone dials the cancel digit 3. Central control, therefore, will release the conference circuit and will establish another talking path in the switching network between customers A and B. However, if the talking paths are busy due to traffic conditions, central control will direct that customers A and B remain connected through the previously established conference circuit.

Another party may be added any time by customer A as previously described. The conference connection is released when customer A disconnects. The usual automatic message account (AMA) records are kept for the use of the conference circuit and the calls involved. Moreover, only one toll connection may be connected to a conference circuit.

## *Dial Conference Service*

A customer can also initiate his own conference calls without the assistance of an operator by means of the dial conference feature provided in electronic central offices. This service, which is limited to one originating customer and three called parties, is initiated when the calling customer A dials the dial conference access code of two digits. This code tells central control that a conference call is to be established for customer A. Accordingly, a dial tone is sent to customer A who dials the add-on digit 2 followed by the directory number of customer B, the first desired party. Central control proceeds to establish a connection between an idle conference circuit and the lines of customers A and B as previously described and illustrated in Fig. 2-18. When customer B answers, he is informed by customer A of the conference call. Customer A now calls the next party C by first flashing his switchhook and, upon receiving a dial tone, dialing the add-on digit 2 and the directory number of customer C. When customer C answers, A repeats the same procedure to add customer D to the conference circuit.

The last party connected to the conference circuit can be removed by the originating customer A flashing his switchhook and when a dial tone is received, dialing the cancel digit 3. All three parties are released when the originating customer A disconnects. Only one toll connection may be made to the conference circuit. The AMA equipment will record the conference circuit usage time and related data for billing purposes in the usual way.

## Administration and Maintenance Features

Supplementing the special customer services that have been described are several unusual administration and maintenance features that are incorporated in electronic switching systems. These capabilities are in addition to the usual testing and maintenance facilities that are also provided in electromechanical offices, such as traffic registers, service observation, local or central test desk connections, station ringer and Touch Tone® set testing. The major items are briefly explained below.

### *Call Tracing*

All incoming calls to a particular customer's line may be traced automatically by incorporating this requirement in the program store memory pertaining to the desired line number. If the originating line number is in the same electronic switching office as the called line that is programmed for call tracing, the teletypewriter at the master control center will print out the originating line number. If the originating line is in another central office, the teletypewriter will print out the equipment number of the associated incoming trunk.

To provide this call tracing feature, it is first necessary that the directory number translator in program store be arranged to require that all incoming calls to the specific line be traced. It is also possible to trace a call in progress to a particular customer's line by requesting such action using the teletypewriter at the maintenance or master control center. In this case, central control will be able to determine the originating line number of the incoming trunk designation from the path memory data in call store.

### *Line Load Control*

In electronic switching offices, customer lines are divided into two general classifications for line load control purposes, namely, essential and nonessential lines. In case of very heavy traffic loads because of storms or other contingencies, a line load control key would be operated at the master control center to give priority service to the essential customers such as police, fire, utilities, or governmental agencies. Other customers will have telephone service to whatever extent is possible, depending upon traffic conditions at any given moment.

### Automatic Overload Control

If excessive delays are encountered on either originating or incoming traffic, such as connecting to idle dial pulse receivers or incoming multi-frequency (MF) tone receivers, lamp and alarm signals at the maintenance control center will alert personnel. At the same time, the teletypewriter will print out a message describing the overload condition, and these printouts will continue as long as the overload lasts.

### Toll Network Protection

The toll network protection feature restricts the use of specified toll trunk groups to toll essential customers. Other customers' calls destined for these toll trunks would be connected to the overflow tone. This protection arrangement is activated by the operation of a special key at the master control center which also activates the line load control operations whether or not that control key is operated. Certain customers that are classified as essential for line load control purposes likewise may be designated as toll essential customers for the toll network protection feature.

### Measurements of Dial Tone Delays

Measuring dial tone delay is an important means for determining the grade of service rendered by a central office. This measurement in electronic switching offices is made by periodic test calls that are automatically initiated. The activation of line load control during overload periods also will start these dial tone delay measurements.

The number of test calls encountering excessive dial tone delays are printed out by the teletypewriter at master control center at periodic intervals. These measurements may be started or stopped at any time by a teletypewriter request.

### Service Order Changes

The temporary memory in call store is also used to record service order changes such as adding or disconnecting customer lines and trunk circuits. A teletypewriter at the business office of the telephone company may be used to transmit the service order number and associated data to the electronic switching central office. The information is recorded in the recent change register in the call store unit. A service order change is made effective when the appropriate code is dialed by operating personnel followed by the specific service order number.

## Questions

1. What are the two main types of electronic switching systems? Which type is used in telephone central offices? Why?
2. What are the designations of the principal electronic switching systems currently developed for city and suburban central office installations? What are the two main parts of these systems? Name three or more major elements associated with the central control unit.
3. What two essential techniques of electronic switching are explained in the analogy to a manual switchboard operated by an electronic operator or automaton? Where is the knowledge stored for the manner of handling calls? What equipment unit does the electronic slate compare to in the electronic switching office?
4. What elements comprise the switching network in an electronic switching office? How many stages involve the complete switching network? What apparatus performs the actual switching operation and how many conductors are switched?
5. What are the functions of trunk, junctor, and service circuits, and with what elements of the electronic switching system are they related? What special type of relay is employed in junctor circuits? Why?
6. What three major classes of instructions normally are processed by central control? Where does central control obtain its processing instructions for handling calls? What other major element is provided in the larger-size No. 1 ESS office to operate with central control, and what is its prime purpose?
7. Define the program store and call store units. How is data or information stored in program store? In call store?
8. What main function is performed by, and how many separate parts comprise the master control center? Name at least three of these parts.
9. What major elements of the electronic switching system are fully duplicated? Name two or more elements which are partially duplicated for system reliability purposes.
10. What three or more special customer services are provided by electronic switching systems that are not available in electromechanical central offices? Name three or more unusual administration and maintenance features incorporated in electronic switching offices.

# 3
# CENTRAL PROCESSOR FUNDAMENTALS

## Elements of Central Processor

In the previous chapter, the basic organization of electronic switching systems was discussed, including the two essential parts, the switching network and the central processor units. Three main elements comprise the central processor: central control, program store, and call store. They govern all system operations in handling telephone calls. In this chapter, the main fundamental functions of the central processor unit, especially central control, will be explained.

Central control may be considered as an electronic information-processing or computer-type device which serves as the center of all call-processing operations. It can request processing instructions from program or call stores, as required for handling particular calls. These requests to program store are sent as an *address* or combination of binary digits which define a specific location in program store's memory. In reply, program store will send back a two-part instruction that had been stored at the particular address. The first part tells central control what actions are to be performed, and the second part, which is the address, indicates where to do it. Central control decodes and executes the received instructions at a rate of one every 5.5 microseconds. This time period is determined by clock pulses that originate in a 2 MHz crystal-controlled oscillator circuit. Figure 3-1 illustrates the fundamental relations between the three elements of the central processor and other components of the electronic switching system.

Three major classes of instructions are usually sent to central control from program store. The first consists of orders to sense the condition or state of lines and trunks—for instance, to check the ferrod line scanners associated with customer lines in order to detect requests for service as indicated by a change of state. In this respect, central control would perceive any change in the state of a customer's line as a binary 0 or binary 1 digit signifying, respectively, off-hook or on-hook conditions. These scanning operations are used to detect inputs to the central processor for necessary actions.

A second class of instructions concerns the processing of the input data. In this connection, central control may manipulate the received data and de-

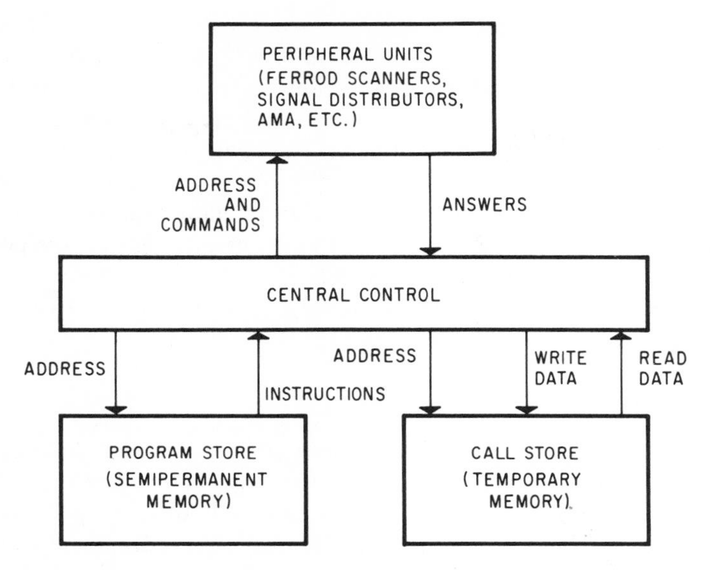


**Fig. 3-1.** Central Processor Elements and Interconnections

posit the results temporarily in call store for later retrieval as may be needed. Additional data also may be obtained from program store as required. These operations do not progress in sequence from one step to the next. If certain conditions are encountered during any stage of the process, central control may decide to transfer to another operation sequence instead of continuing with the particular program.

The third class of instructions pertains to generating outputs from central control to operate relays in junctor, service, and trunk circuits, to close ferreed switches in line-link, trunk-link, or other network paths, etc. A single instruction usually controls one operation. However, individual instructions can be combined in different desired ways for implementing any combination of control requirements.

The information and instructions deposited in program store's memory are arranged in binary words. Each word consists of a combination of 44 binary digits or bits. A binary digit can be either 0 or 1. Every binary word occupies a location in the program store's semipermanent memory that is uniquely identified by an address. Thus, when a particular address is received from central control, program store will send back (read out) the corresponding binary word to central control. The memory in program store is termed semipermanent because it can be changed only by external means.

Information is deposited in call store in a similar manner except that the binary words used contain only 24 binary digits or bits. Inputs or commands from central control to call store specify, in addition to the address or word location, the type of operation to be performed. For example, to interrogate (read) or determine the state of a particular circuit, to delete previously written data, or to insert (write) information. In the case of an interrogation (read) request, call store will transmit (read out) to central control the binary word at the indicated address. Call store contains a temporary memory because it can be changed by central control at any time.

The binary digit or bit (as it is usually abbreviated) refers to characters in the binary number system. A binary digit actually represents one of two possible conditions or logic states, such as on or off, high or low potential, magnetized or demagnetized, and conducting or not conducting. These logic states are symbolized by the binary digit 0 or 1 in the binary or two-number system.

## Binary Number System

The decimal system in common universal use is based on ten possible digits, 1,2,3,4,5,6,7,8,9, and 0, and the mathematical powers of 10 or $10^n$. For instance, any one of the ten digits may be entered in the first space to the left of a decimal point. This digit, consequently, would be multiplied by ten every time it is moved another position to the left, for example, 7, 70, 700, 7000, 70,000, 700,000, etc.

The binary system, on the other hand, has only two digits, 0 and 1. And it operates on the mathematical powers of 2 ($2^n$) in contrast to the decimal system's $10^n$ powers. It is possible, of course, to have number systems based on other digits, such as, $3^n$, $4^n$, $5^n$, $6^n$, etc. The binary or two-digit numbering system was chosen for the basis of the machine language used by computers and similar devices because only two digits, 0 and 1, are needed to express a logic state or condition encountered, for instance, true or false, on or off, yes or no, etc.

Numbers in our decimal system may be converted to binary form and vice versa by the use of binary coded decimal tables. One form of these tables consists of columns extending from right to left with each column having a value or weight based on the powers of 2, except that the first column on the extreme right always will have a value of 1. For example, in a seven-column table, the weights reading from right to left would be: 1, $2^1=2$, $2^2=4$, $2^3=8$, $2^4=16$, $2^5=32$, and $2^6=64$.

A 4-bit numbering arrangement used in some electronic switching elements is shown in Table 3-1. Each decimal number from 1 to 15 is represented by a binary number of four digits. Note that only binary digit 1 is assigned a weight or value, based on the powers of 2, in each of the four columns. These values start with 1 for column D at the extreme right, 2 for column C, 4 for column B, and 8 for column A. The summation of columns A+B+C+D, wherever binary digit 1 appears, will give the corresponding decimal number. For instance, decimal number 4 is composed of binary digits 0100 in that order. The decimal number equivalent, 4, is obtained from the value or weight

**Table 3-1.** Binary Number System for Decimal Digits 0–15

| Decimal Number | Four-digit Binary Code A (8) | B (4) | C (2) | D (1) |
|---|---|---|---|---|
| 0 | 0 | 0 | 0 | 0 |
| 1 | 0 | 0 | 0 | 1 |
| 2 | 0 | 0 | 1 | 0 |
| 3 | 0 | 0 | 1 | 1 |
| 4 | 0 | 1 | 0 | 0 |
| 5 | 0 | 1 | 0 | 1 |
| 6 | 0 | 1 | 1 | 0 |
| 7 | 0 | 1 | 1 | 1 |
| 8 | 1 | 0 | 0 | 0 |
| 9 | 1 | 0 | 0 | 1 |
| 10 | 1 | 0 | 1 | 0 |
| 11 | 1 | 0 | 1 | 1 |
| 12 | 1 | 1 | 0 | 0 |
| 13 | 1 | 1 | 0 | 1 |
| 14 | 1 | 1 | 1 | 0 |
| 15 | 1 | 1 | 1 | 1 |

in column B in which binary digit 1 appears. In a similar fashion, binary number 0001, which has a value of 1 in column D, corresponds to decimal number 1, and binary number 0101 which has a value of 4+1 (per columns B and D) will be analogous to decimal number 5.

For decimal numbers between 16 and 31, it will be necessary to utilize five binary digits with values of 16,8,4,2, and 1, respectively, for each binary digit 1 as shown in Table 3-2. Since the summation of 16+8+4+2+1 = 31,

**Table 3.2.** Binary Number System for Decimal Digits 16-31

| Decimal Number | Five-digit Binary Code (16) | (8) | (4) | (2) | (1) |
|---|---|---|---|---|---|
| 16 | 1 | 0 | 0 | 0 | 0 |
| 17 | 1 | 0 | 0 | 0 | 1 |
| 18 | 1 | 0 | 0 | 1 | 0 |
| 19 | 1 | 0 | 0 | 1 | 1 |
| 20 | 1 | 0 | 1 | 0 | 0 |
| 21 | 1 | 0 | 1 | 0 | 1 |
| 22 | 1 | 0 | 1 | 1 | 0 |
| 23 | 1 | 0 | 1 | 1 | 1 |
| 24 | 1 | 1 | 0 | 0 | 0 |
| 25 | 1 | 1 | 0 | 0 | 1 |
| 26 | 1 | 1 | 0 | 1 | 0 |
| 27 | 1 | 1 | 0 | 1 | 1 |
| 28 | 1 | 1 | 1 | 0 | 0 |
| 29 | 1 | 1 | 1 | 0 | 1 |
| 30 | 1 | 1 | 1 | 1 | 0 |
| 31 | 1 | 1 | 1 | 1 | 1 |

six binary digits would be needed to represent decimal numbers in the sequence from 32 to 63 inclusive, and seven binary digits will be required for decimal numbers 64 to 127 inclusive. For example, binary digits 110010 represent decimal number 50, 1100100 corresponds to 100, and 1111111 to decimal number 127.

It is also possible to assign other binary code weights or values for the decimal digits. These values do not necessarily have to follow the $2^n$ mathematical sequence. For instance, a four-digit binary code may have values of 5421, 6421, 7421, etc. Moreover, a 2-out-of-5 code for error detecting purposes may be utilized in certain elements of the central processor. A five-digit binary code for this purpose could have weights of 74210 for binary digit 1 in its respective five columns.

## Binary Logic Principles and Circuits

The binary number system also may be considered as a form of mathematics applicable to binary logic circuits. This follows because any circuit producing on-off or equivalent type of operations possesses binary logic. And any system or subsystem having inputs and outputs which are mainly on and off operations likewise utilizes binary logic principles.

Electronic switching systems require large numbers of logic circuits, especially electronic gates, which can operate exceedingly fast and share a common control unit. An electrical circuit can be described as performing logic if it is capable of generating a high (binary 1) output only when certain specified input combinations exist. For instance, assuming two inputs, the output would be high only if the two inputs are equal, that is, both are high or both are low. The logic circuit compares the two inputs and provides a high output only when both inputs are equal.

The operation of electronic gates including the low-level logic circuits widely employed in central control may be better understood by first considering the basic types of logic circuits involved. Basic logic circuits or modules are designated AND, OR, and NOT gates in accordance with the functions performed. Modules may be combined to form other logic circuits, such as NAND (NOT-AND) and NOR (NOT-OR) gates. These fundamental logic functions and related circuit examples are explained below with respect to their use in electronic switching systems.

### *Logic AND Gate*

To understand logic functions, let us consider a simple electrical circuit consisting of two switches in series that control a lamp as illustrated in Fig. 3-2. The lamp (output) will light (function) only if both switches A and B are closed to connect the power source. This particular electrical circuit is a form of a logic AND gate because both switches A *and* B must be closed for the

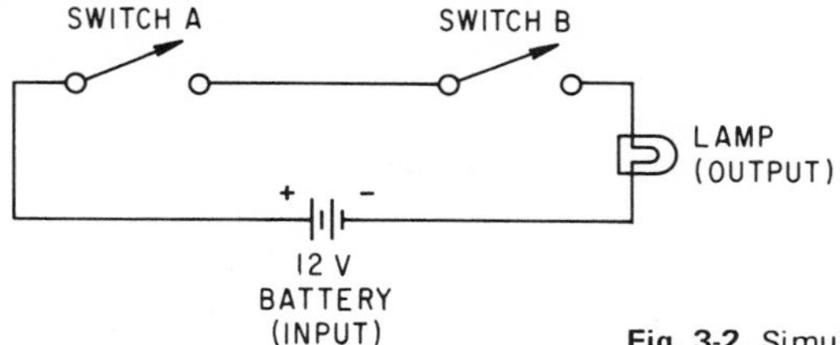


**Fig. 3-2.** Simulated Circuit for AND Logic Function

output to function. The switches and the lamp in this circuit may be replaced with diodes, transistors, or relays to give the same AND logic functions.

The logic symbol for the AND gate is shown in Fig. 3-3. Since there are only two possible states for switches A and B, either on or off, the binary number system is employed to denote the condition or state of all possible circuit combinations of switches A and B. Using positive logic, binary digit 1 is normally used to represent the "on" or "high" state and binary digit 0 the "off" or "low" state for any input or output of the logic circuit. All possible combinations of inputs to a logic circuit and the resultant outputs, such as for the AND gate in Fig. 3-3, may be presented by a tabulation of the respective binary digits in a format called a Truth Table.

The Truth Table for the simple AND gate is shown in Table 3-3. Columns A and B of this table list the binary digits representing the states of the inputs, for example, whether switches A and B in Fig. 3-2 are open or closed. Column C indicates the binary digit outputs of this AND gate. Referring to the basic AND logic circuit in Fig. 3-2, it is seen that when both switches are open, the resultant input state is represented by binary digit 0 in the first row of columns A and B of the Truth Table. Since no current flows through the lamp, the corresponding output in the table's first row of column C likewise will be binary digit 0. When only switch A is closed, column A in the second row will have a binary digit 1 but columns B and C will remain binary 0. Similarly, if only switch B is closed, column B in the third row will show a binary 1 but columns A and C will be binary 0. However, when switches A and B are both closed, current will flow to light the lamp. Therefore, the fourth or last row in this Truth Table will have binary digit 1 in columns A, B, and C.

The corresponding relations in the Truth Table also may be expressed by the following mathematical formula (with the dot meaning logic AND) called a Boolean equation:

$$C = AB \text{ or } C = A \cdot B$$

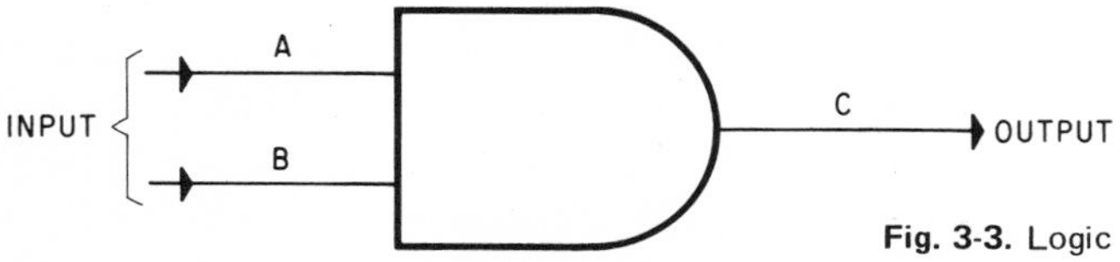

INPUT A B C OUTPUT

**Fig. 3-3.** Logic Symbol for AND Gate

**Table 3-3.** Truth Table for Basic AND Gate

| Inputs | | Outputs |
|---|---|---|
| A | B | C |
| 0 | 0 | 0 |
| 1 | 0 | 0 |
| 0 | 1 | 0 |
| 1 | 1 | 1 |

This kind of mathematics is named after George Boole who, in 1847, introduced a mathematical form of shorthand to represent the functions of logic systems. It is not within the scope of this book to attempt to explain Boolean theorems and equations. Consequently, only the simple fundamental equations for basic logic circuits will be utilized as may be required.

## *Logic OR Gate*

The logic OR gate, as its name implies, signifies that the output will respond to either of two separate inputs or to both. This particular logic function may be described by the lamp circuit schematic in Fig. 3-4. This schematic is a rearrangement of the switches in the AND logic circuit that was illustrated in Fig. 3-2. It will be apparent from Fig. 3-4 that the lamp will light when either switch A *or* B is closed, or if both switches are closed; hence the name logic OR gate. This type of logic circuit is also known as the INCLUSIVE-OR gate.

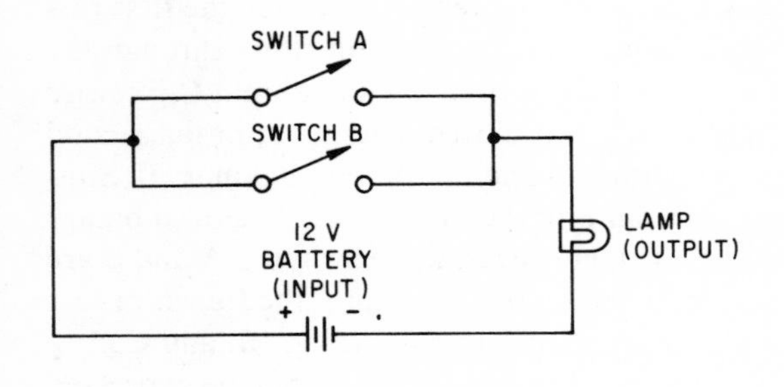


**Fig. 3-4.** Simulated Circuit for OR Logic Function

The usual logic symbol for the OR gate is depicted in Fig. 3-5 and its corresponding Truth Table is listed in Table 3-4. Note that the curved line between inputs A and B in Fig. 3-5 is used to indicate an OR logic function in contrast to the straight line between inputs in Fig. 3-3 for the AND logic gate.

When both switches A and B in Fig. 3-4 are open, current will not flow to light the lamp. Therefore, binary 0 digits are entered in columns A, B, and C of the first row in the corresponding Truth Table per Table 3-4. If only switch A is closed, the second row will have a binary 1 in the A column, binary 0 in the B column (switch B is open), and a binary 1 in column C since the

Fig. 3-5. Logic Symbol for OR Gate

Table 3-4. Truth Table for OR Gate

| Inputs | | Outputs |
|---|---|---|
| A | B | C |
| 0 | 0 | 0 |
| 1 | 0 | 1 |
| 0 | 1 | 1 |
| 1 | 1 | 1 |

lamp lights. For the same reasons, if only switch B is closed, the third row of the Truth Table will indicate 011 for columns A, B, and C, respectively. When both switches A and B are closed, the Truth Table will have binary 1 digits in all columns A, B, and C as shown in the fourth or bottom row.

The corresponding Boolean equation for a logic OR gate is usually written as: $C = A+B$. This signifies that the output of an OR logic gate is equal to either A or B.

## *Inverter NOT Gate*

The simulated logic circuits drawn in Figs. 3-2 and 3-4 may be replaced by diodes and associated components for electronic circuit applications. Diodes are passive elements and can cause attenuation to input signals. The output signal level, therefore, may not be sufficient to drive a network of additional electronic gates or other circuits. Consequently, a transistor of the common-emitter type often is connected to the outputs of the diodes in order to amplify the signals to the required level. The transistor's output signal, however, will be inverted with respect to its input signal state. That is, a binary digit 1 at the input to the transistor becomes inverted to a binary digit 0 at the output. Similarly, a binary 0 at the transistor's input changes to a binary 1 at its output.

This inversion action of the transistor results in a NOT logic gate. The logic symbol for a NOT gate is shown in Fig. 3-6. The triangle in this figure denotes the amplification action of the transistor. The small circle at the apex of the triangle indicates that the output is reversed with respect to the input signal. For example, if the input is a binary 1, the output is a binary 0. For simplicity in logic symbols, it is customary to omit the triangle and to represent a transistor in a logic circuit by a small circle at the output end of the particular circuit's logic symbol. This method is used in Fig. 3-7 for the NAND gate's logic symbol.

Fig. 3-6. Logic Symbol for Inverter or NOT Gate

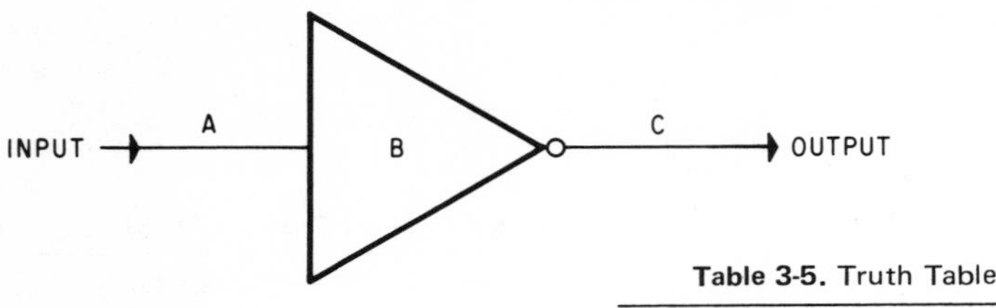


Table 3-5. Truth Table for NOT Gate

| Input | Output |
|---|---|
| A | C |
| 1 | 0 |
| 0 | 1 |

The Truth Table for the inverter or NOT gate in Table 3-5 is very simple because there are only two possible inputs and outputs. The Boolean equation for this logic function is usually written as: $C = \overline{A}$. This expression means that the output is *not* the input because a bar or line over a letter signifies "not" in Boolean mathematical usage.

## *Logic NAND (NOT-AND) Gate*

Combining the logic AND gate of Fig. 3-3 with a transistor (inverter) is equivalent to combining the AND gate with a NOT gate as in Fig. 3-6. The result will be a logic NAND gate also known as the NOT-AND or AND-NOT gate. Figure 3-7 is the logic symbol for the NAND gate, and its Truth Table is listed in Table 3-6.

The output of the NAND gate, per column C of its Truth Table, will be negative (binary 0) only if both inputs A and B are positive (binary 1). Note

Fig. 3-7. Logic Symbol for NAND (NOT-AND) Gate

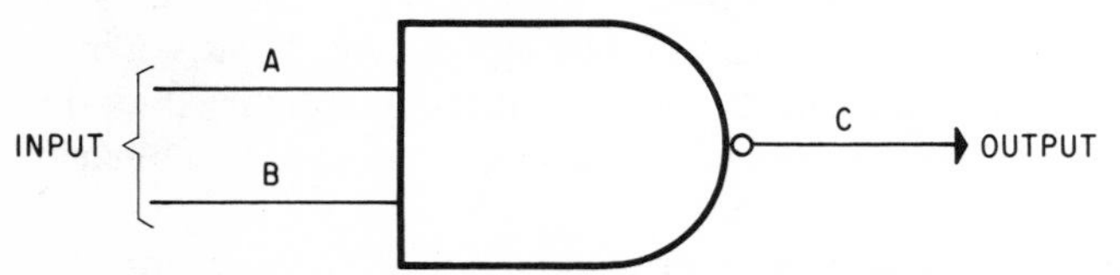


Table 3-6. Truth Table for NAND Gate

| Inputs | | Outputs |
|---|---|---|
| A | B | C |
| 0 | 0 | 1 |
| 1 | 0 | 1 |
| 0 | 1 | 1 |
| 1 | 1 | 0 |

that columns A and B of the Truth Table for the AND gate (Table 3-3) correspond to columns A and B of the Truth Table for the NAND gate in Table 3-6. The outputs listed in column C of the NAND gate will be reversed as compared to the AND gate because of the transistor's inversion action.

## *Logic NOR (NOT-OR) Gate*

The logic NOR gate, which also is known as a NOT-OR or OR-NOT gate, is the result of combining a logic OR gate with a transistor or inverter. The usual logic symbol for a NOR gate is portrayed in Fig. 3-8. It is similar to the OR gate symbol in Fig. 3-5 with the addition of the small circle at the output to indicate the inversion action of its transistor.

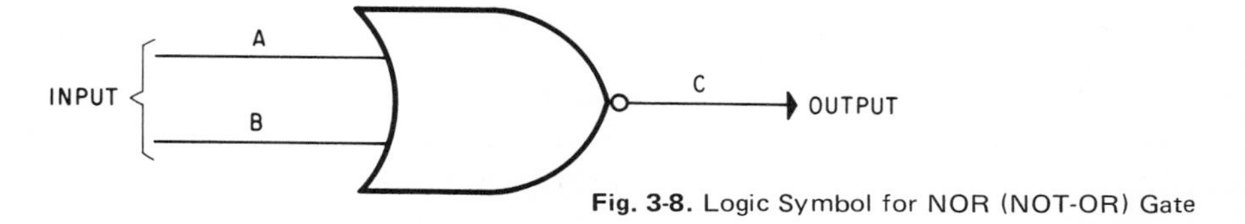


**Fig. 3-8.** Logic Symbol for NOR (NOT-OR) Gate

When separate transistors are required for each input, as is often the case in many electronic switching circuits employing low-level logic, the logic symbols in Fig. 3-9 are usually employed for the NOR gate. In either case, the output of the NOR gate will be positive (binary 1) only if both inputs to the NOR gate are negative (binary 0).

**Fig. 3-9.** NAND Gates Used as a NOR Gate

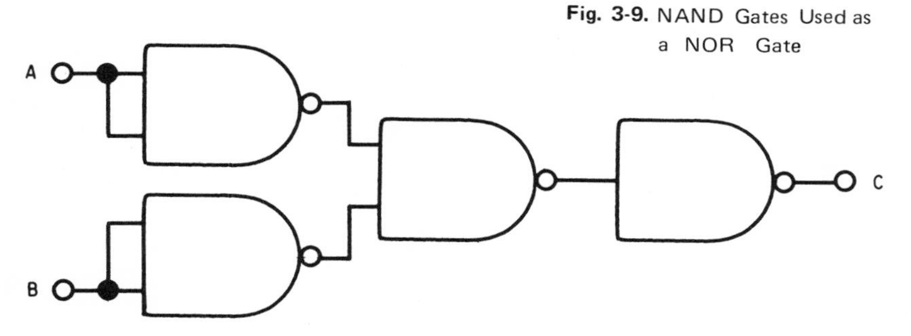


**Table 3-7.** Truth Table for NOR Gate

| Inputs | | Outputs |
|---|---|---|
| A | B | C |
| 0 | 0 | 1 |
| 1 | 0 | 0 |
| 0 | 1 | 0 |
| 1 | 1 | 0 |

The Truth Table for the NOR gate appears in Table 3-7. Note that the binary digits in columns A and B are the same as the respective columns of the OR gate's Truth Table in Table 3-4. However, the binary outputs in column C of the NOR gate's Truth Table (Table 3-7) are inverted with respect to those in column C of the OR gate's Truth Table in Table 3-4 because of the transistor.

The Boolean equation for a NOR gate may be written as:

$$C = (\bar{A}+\bar{B}) \quad \text{or} \quad C = (A \cdot \bar{B}) + (\bar{A} \cdot B)$$

The plus sign signifies logic OR, the dot designates logic AND, and the bar or line over the letters means "not," as previously explained.

## *Low-Level Logic Circuit*

The basic logic gates that have been described can be merged into various circuit combinations for the different electronic switching requirements. Although, for simplicity, the inputs to the logic gates in the different referenced figures have been limited to two, many more inputs may be added as required. The same rules of logic will still prevail.

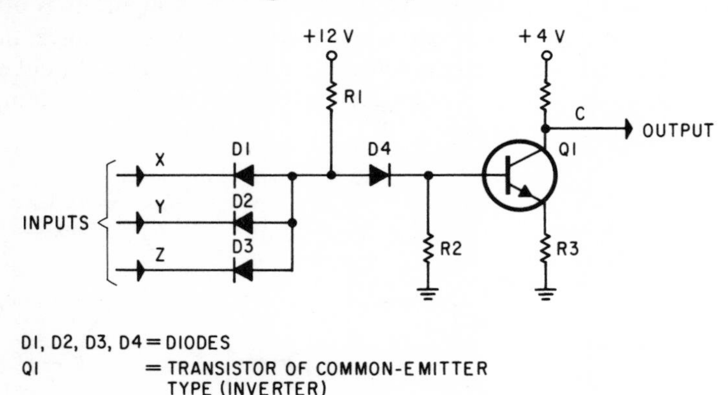


**Fig. 3-10.** Low-level Logic NAND (NOT-AND) Gate

One particular logic combination, employed in central control and other electronic switching elements, is the low-level logic circuit illustrated in Fig. 3-10. It consists of a NAND or AND-NOT gate that incorporates a voltage limiting or bias network for the transistor. This bias circuit, indicated in Fig. 3-10 by diode D4 and resistor R2, provides large reverse bias drive currents for transistor Q1. The arrangement permits very fast switching speeds in this particular logic circuit.

## Logic Circuit Applications

The importance of logic circuits in central control may be exemplified by the line scanning functions performed by the ferrod line scanner illustrated in Fig. 2-9. This circuit is directed by central control to scan its related customer lines at least once every 200 milliseconds in order to ascertain whether or not a change of state has occurred. The application of logic circuits utilizing NAND, OR, and AND gates for this purpose is explained in the following example:

Assume that central control is to determine whether or not a change of state has taken place for line 2345. Central control will direct call store to look in the address of line 2345 and read out its last recorded state. At the same time central control will obtain the present state of this line from the associated ferrod line scanner circuit. Referring to Fig. 2-9, recall that if line 2345 is on-hook, its ferrod sensor will produce a readout pulse, and a binary digit 1 will be returned to central control. If the line is off-hook, no pulse will be generated during the scanning process, and a binary digit 0 will be sent back to central control as an indication of the line's present state.

The last and present states of line 2345 are received simultaneously by the specified logic circuit in central control consisting of logic gates OR, NAND, and AND as illustrated in Fig. 3-11. If line 2345 is off-hook, a binary digit 0 will be transmitted by the ferrod line scanner to the A input of the OR and NAND logic gates. Similarly, if the last state of this line was recorded as being off-hook, call store will send a binary 0 to the B input of the OR

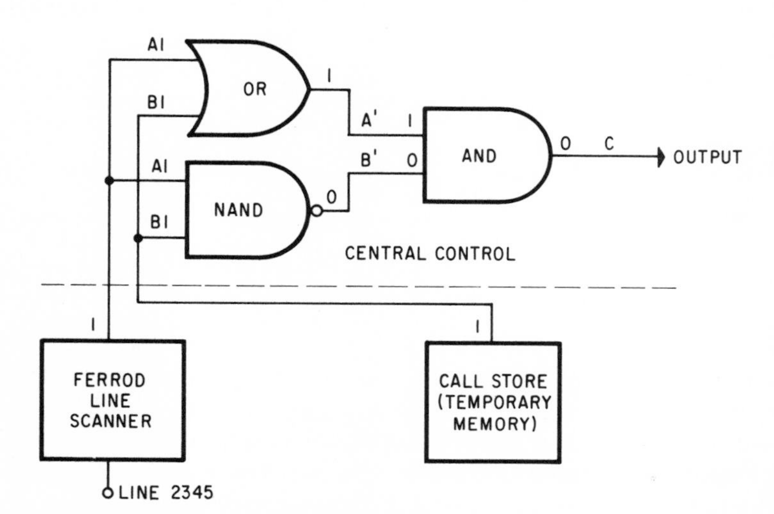


**Fig. 3-11.** Logic Symbol for OR, NAND, and AND Gates and EXCLUSIVE OR (EXC OR) Logic Circuit

and NAND gates. Referring to the Truth Table for an OR logic gate in Table 3-4, it is noted that inputs of binary 0 will generate a binary 0 output. Thus, a binary digit 0 will be present at the A′ input of the AND gate. At the same time, the two binary 0 inputs to the NAND gate will, in accordance with the Truth Table in Table 3-6, produce a binary 1 output. Therefore, a binary 1 will be transmitted to the B′ input of the AND logic gate. The resultant output C of the AND gate will be a binary 0, in accordance with the Truth Table in Table 3-3, which will indicate to central control that there has been no change of state in line 2345.

If both the last and present states of line 2345 were on-hook, there would be binary digits 1 on the respective A and B inputs of the OR and NAND gates in Fig. 3-11. The output C of the AND gate likewise will be binary 0 since the condition or state of line 2345 has not changed. This condition is represented by the binary digits on the different logic gates in Fig. 3-11, which also represents an EXCLUSIVE OR (EXC OR) logic circuit.

Now suppose that the ferrod line scanner circuit indicates by a binary 0 that line 2345 is off-hook but that call store reports by a binary 1 signal that the last state was on-hook. The respective A and B inputs to the OR and NAND logic gates in this case will be binary 0 and 1, respectively. The outputs of the OR and NAND gates will both be binary 1 in accordance with their respective Truth Tables per Tables 3-4 and 3-6. The application of binary 1 digits to the A′ and B′ inputs of the AND gate will result in a binary 1 output in central control, thereby indicating that a change of state has occurred for line 2345. These various customer line conditions and the corresponding binary signals are tabulated in Table 3-8 for reference purposes.

**Table 3-8.** Customer Line Conditions and Resultant Binary Signals

| Line Scanner Circuit (Present State) | Call Store (Last State) | Output Signal in Central Control | Significance |
|---|---|---|---|
| 0 | 0 | 0 | State not changed (Off-hook) |
| 1 | 1 | 0 | State not changed (On-hook) |
| 0 | 1 | 1 | State changed (Off-hook) |
| 1 | 0 | 1 | State changed (On-hook) |

## Basic Flip-Flop Memory Circuits

Groups of logic circuits may be combined to provide for temporary storage of information in the form of binary digits or bits of information. One particular form is the flip-flop arrangement that is extensively employed as a memory circuit in central control. An electrical circuit is said to possess memory if, after being placed in a certain state by an input signal, it remains in that state after the removal of the input signal. For example, a latching relay that operates and locks up has memory because it does not release when the operating current is removed.

Two NAND or AND-NOT gates are often combined for a flip-flop circuit although two NOR or OR-NOT gates also may be utilized. The resultant flip-flop circuits are in Figs. 3-12 A and B. The flip-flop nomenclature originated from the use of these circuits in computers and related equipments. The two stable states are usually called set and reset and abbreviated S and R, respectively. The flip-flop action permits the storage of information in binary form as will be explained. Flip-flop circuits remain in either state until an input signal or associated timing or clock pulses cause a change to the other state.

In central control, the usual flip-flop circuit consists of two NAND logic gates of the low-level type as shown in Fig. 3-10. The block diagram and logic symbol are shown in Table 3-9. The two inputs to this flip-flop are commonly designated S and R for "set" and "reset," and the two outputs by the

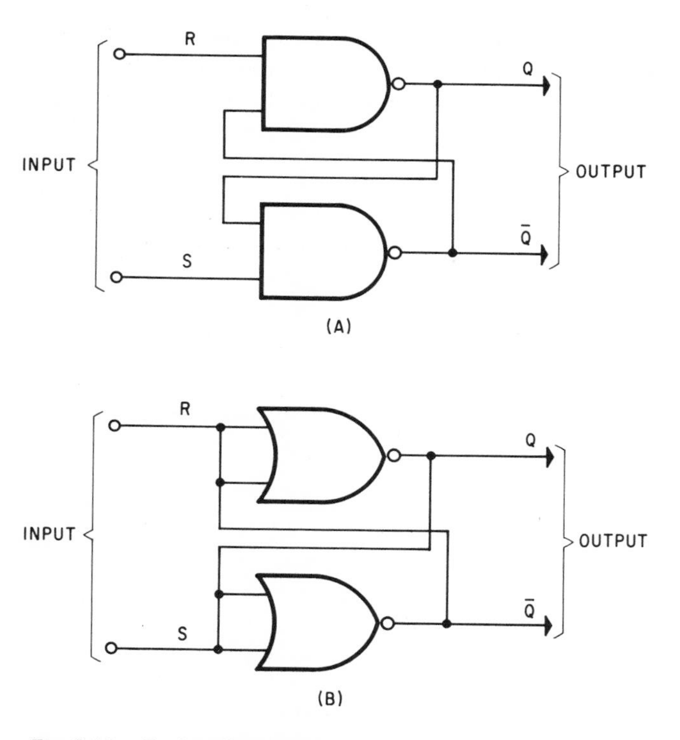


**Fig. 3-12.** Typical Flip-flop Circuits A: Two NAND Gates
B: Two NOR Gates

**Table 3-9.** Basic Flip-flop Truth Table

| Inputs | | Outputs | |
|---|---|---|---|
| S | R | Q | $\overline{Q}$ |
| 1 | 0 | 1 | 0 |
| 0 | 1 | 0 | 1 |
| 0 | 0 | No Change | |
| 1 | 1 | Undetermined | |

letters Q and $\overline{Q}$. The bar or line over $\overline{Q}$ means "not Q" as was explained for the logic NOT and NAND gates. In view of the importance of flip-flop circuits for recording information in central control and other major elements of the central processor, a brief and simplified explanation of flip-flop logic operations follows.

Referring to Fig. 3-13, a brief synopsis of the flip-flop circuit operations is that inputs to R and S normally are in the "high" state or binary 1 whenever the flip-flop is inactive. This results because when the set (S) input is high or binary 1, the Q output will be 1 and the Q output will be 0. Consequently, any change in the state of the R input will not affect the output of $\overline{Q}$ because the cross-connection from $\overline{Q}$ to an input of the NAND 1 gate will keep the transistor in NAND 1 nonconducting. Now, if the input to R becomes binary 0, the input to the transistor of NAND 2 will be decreased and its $\overline{Q}$ output will change from binary 0 to 1. This reversal of $\overline{Q}$ from a low to

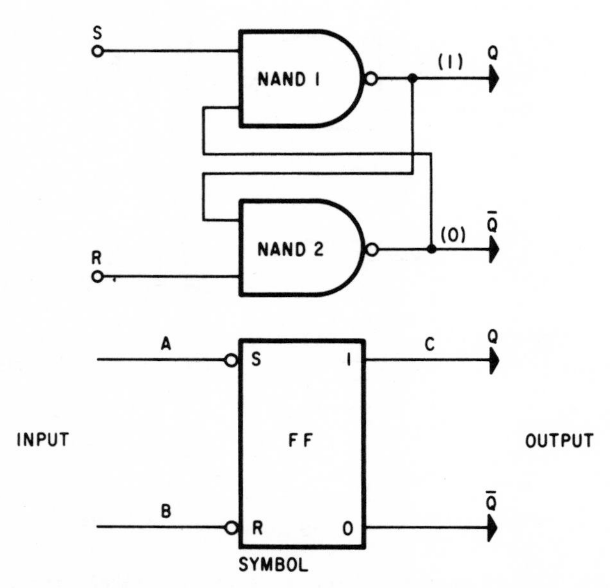


**Fig. 3-13.** Low-level Logic Flip-flop Circuit

high state will increase the drive current to the transistor in NAND 1 so that it will conduct. The Q output, consequently, will change from binary 1 to 0. These actions will continue in unison with changes in the respective input states.

## J-K Flip-Flop Logic Circuit

The basic flip-flop circuit has some disadvantages especially when the inputs are simultaneously binary digits 1. This situation may be understood by examining the flip-flop Truth Table in Table 3-9. Observe that the Q output is a function of the flip-flop's outputs and its inputs at the previous time of the binary digit. For instance, if the set (S) input is binary 1 and the reset (R) input is binary 0, the Q output will be binary 1 as shown in the first row of the Truth Table. If the S and R input states are reversed, the Q and $\overline{Q}$ output states likewise will be reversed as indicated in the second row of the Truth Table by the respective binary digits. However, when the inputs are both binary 0, the output will not change from what it was in the previous bit time. Also, when the inputs are simultaneously binary digits 1, the next state cannot be determined.

To overcome this undesirable flip-flop situation and to provide for binary counting or division operations, use is made of the J-K flip-flop circuit. This device, which can handle all possible input combinations, is represented by the symbol in Fig. 3-14. This symbol comprises two basic flip-flops in series. The inputs marked J and K influence the flip-flop operations only when synchronized with the clock or timing pulses that control the operations. The set and reset inputs operate directly on the output without being synchronized with the clock pulses.

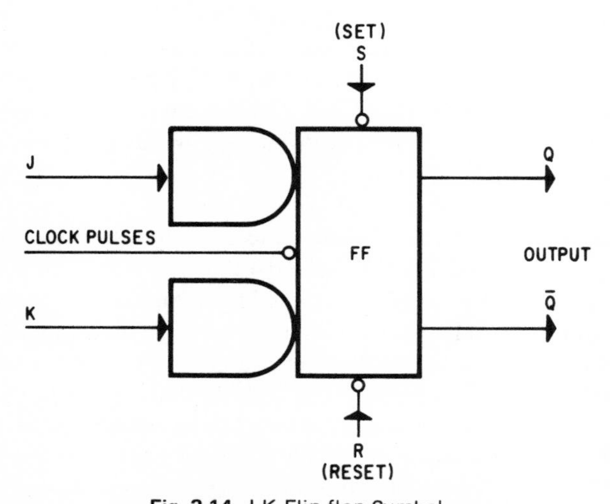


**Fig. 3-14.** J-K Flip-flop Symbol

**Table 3-10.** J-K Truth Table

| Inputs | | Outputs | |
|---|---|---|---|
| J | K | Q | $\overline{Q}$ |
| 1 | 0 | 1 | 0 |
| 0 | 1 | 0 | 1 |
| 0 | 0 | No Change from Previous State | |
| 1 | 1 | Changes to Opposite State | |

In the J-K flip-flop memory circuit, the Q output will change to the opposite state when both J and K inputs are binary 1. The output will not change if both J and K inputs are binary 0. When a binary 0 is applied at the reset input, the output of the J-K flip-flop will be in the zero-state position or Q = 0. If a binary 0 is applied to the set input, the J-K flip-flop will be placed in the set state or Q = 1. The Truth Table in Table 3-10 illustrates these fundamental relations which are specifically applicable to the storing of information in central control.

Various combinations of flip-flop memory circuits are utilized in central control and other elements of the central processor. These include different types of registers, buffers, binary counters, and associated units. The major types and their functions will be described.

## Register Circuit

A major use of flip-flops is to form a register circuit which has the capacity to store one binary word. One flip-flop is needed for each binary digit or bit comprising the stored binary word. For the registers used in central control, binary words normally consist of 23 bits including one bit used to represent the sign of the word. This sign bit is 0 for positive numbers and 1 for negative numbers. Approximately ten 23-bit registers are usually provided in central control for temporarily storing binary words or information during call processing operations.

A typical register in central control may be represented by the simplified block diagram of the NAND logic gates and flip-flop memory symbols in Fig. 3-15A. Recall from Fig. 3-10 that the output of a NAND gate will be high (+4 volts) when at least one of the inputs is low (0 volts), and that its output will be low only when all inputs are high. This nomenclature is used in the following explanation of the operation of the register circuit.

We will assume, referring to Fig. 3-15A, that all flip-flops in the register circuit have been reset by a momentary low (0 volts) pulse on the common reset lead that is connected to all R inputs. When a high (binary 1) is received on gate lead A, the binary information on input leads $A_0$ to $A_{22}$ inclusive will be sent to the corresponding 23 flip-flops designated FF BIT 0 to BIT 22.

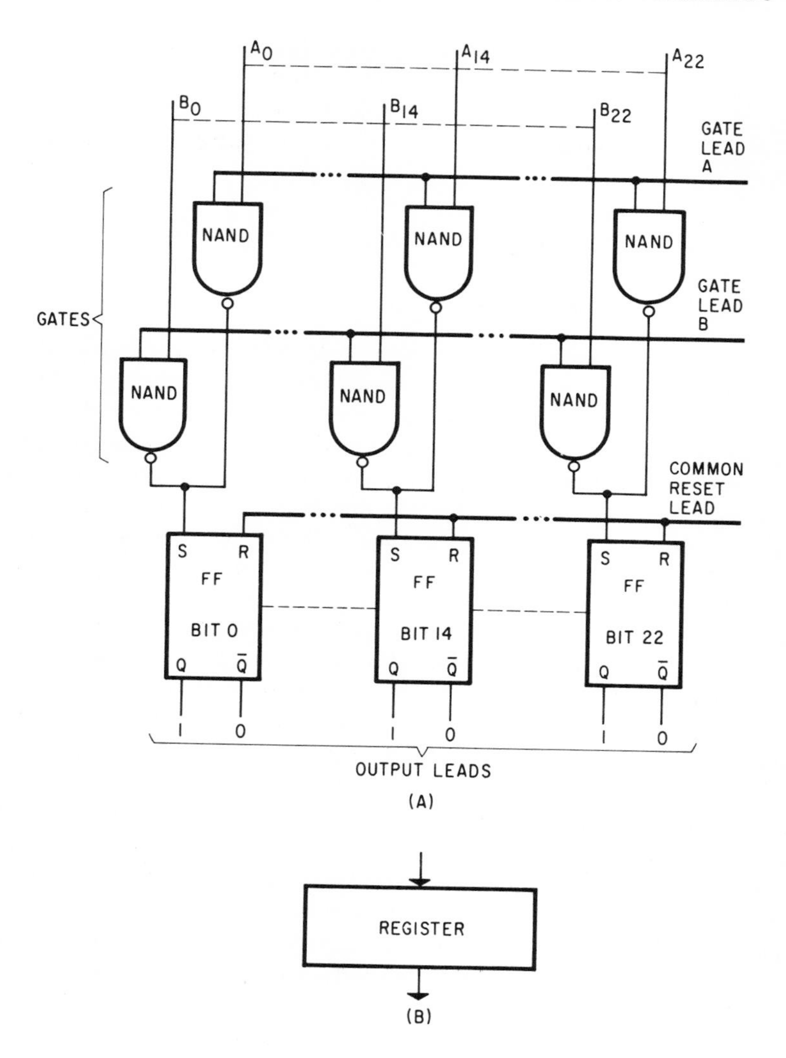


**Fig. 3-15.** Block Diagram of a 23-Bit Register and Its Symbol

Note that each flip-flop is marked by its related bit number to simplify the description of the operation of the register.

For example, if the input to lead $A_{14}$ should be high, when gate A closes, flip-flop BIT 14 remains in its reset state. The same conditions apply to the other input leads in the $A_0$ to $A_{22}$ sequence. In a similar manner, when gate lead B is high, the binary information on input leads $B_0$ to $B_{22}$ inclusive will be forwarded to their respective flip-flops. Thus, gates A and B determine when the input data is to be sent to the register and from what source. For instance, gate A may control the inputs from program store and gate B from scanner circuits.

The symbolic form of a register is shown in Fig. 3-15B. The control gates are often omitted for simplicity purposes. The different registers provided within central control are usually designated by the functions that they perform. For example, L for logic register, F for "first one" register, K for accumulator register, and the X, Y, and Z index registers.

## Program Instruction Fundamentals

Before proceeding with the descriptions of the organization and operation of central control, it is desirable that the principal concepts of program instructions and operations associated with them be understood. The stored program in electron switching offices utilizes a large number of various types of instructions. The program is organized so that an interrupt system initiates the input-output program instructions that must be performed in accordance with an accurate timing tolerance. Data collected by these input-output programs are forwarded to the call processing programs which decide the course of action for handling calls.

Three basic types of program instructions generally are employed. They may be identified by the functions performed, for example, the *perform type*, the *proceed type*, and the *examine type*. A perform type of instruction would request central control to carry out a specific operation and then advance directly to the next instruction. A proceed type would require central control to go unconditionally to a specified address in the program and to proceed from that point. An examine instruction would require central control first to determine the value of some specified binary quantities and then to decide whether to advance to the next instruction or to transfer the data to a specified address in the program. The decision to transfer is based on the sign and *homogeneity* of a particular binary word. In this respect, recall that one of the 23 bits comprising a binary data word represents the sign of the word, and that this sign bit is 0 for positive numbers and 1 for negative numbers. The homogeneity, which may be defined as the mathematical state of a data word, is 1 if *all* of its bits are either 1 or 0; otherwise homogeneity is 0.

The fundamental structure of a typical program incorporating the aforementioned types of instruction is outlined in Table 3-11. The actual program for an electronic switching office employs many instructions of various types and has a very complex pattern. However, it is similar in concept to the program shown in Table 3-11. For instance, the binary number sequence and its equivalent decimal number represent the address or location of a specific instruction in program store. Each of the letters in the horizontal blocks symbolizes a particular type of instruction. For example, in the first column, instructions A and B are of the same perform type, each advancing without restrictions to the next address. Instruction C is the examine type which would

Table 3-11. Fundamental Structure of a Program

| 1 | | | 2 | | | 3 | | |
|---|---|---|---|---|---|---|---|---|
| A* | B† | C‡ | A | B | C | A | B | C |
| (0) | 00000 | A | | | | | | |
| (1) | 00001 | B | | | | | | |
| (2) | 00010 | C | | | | | | |
| (3) | 00011 | D | (16) | 10000 | Q | | | |
| (4) | 00100 | E | (17) | 10001 | R | | | |
| (5) | 00101 | F | (18) | 10010 | S | | | |
| (6) | 00110 | G | (19) | 10011 | T | | | |
| (7) | 00111 | H | | | | | | |
| (8) | 01000 | I | | | | | | |
| (9) | 01001 | J | | | | | | |
| (10) | 01010 | K | (20) | 10100 | U | (24) | 11000 | Y |
| | | | (21) | 10101 | V | (25) | 11001 | Z |
| (11) | 01011 | L | | | | | | |
| (12) | 01100 | M | (22) | 10110 | W | (26) | 11010 | AA |
| (13) | 01101 | N | (23) | 10111 | X | (27) | 11011 | AB |
| (14) | 01110 | O | | | | (28) | 11100 | AC |
| | | | | | | (29) | 11101 | AD |
| (15) | 01111 | P | | | | (30) | 11110 | AE |

* Equivalent decimal number
† Address in binary numbers
‡ Type and text of instruction

direct central control either to instruction D or to instruction Q. Instructions D to O, inclusive, also are of the perform type leading unconditionally to the next address.

Referring to column 2, instructions Q, R, and S are perform-type instructions and T is an examine type leading either to instruction U or Y. Instructions U, V, and W are all of the perform type. Instruction X is of the proceed type which leads directly to instruction L. Instructions Y, Z, AA, AB, AC, and AD in column 3 likewise are of the perform type. Instruction AE is a proceed type leading directly to instruction P in column 1. Instruction P also is a proceed type which leads back to the initial instruction A.

## Program Instruction Components

Program instructions, such as those received from program store, may be divided into four parts or fields. These parts are usually designated *operation* (OP), *indentification constant, data address* (DA), and the *options field.* Program instructions are written mnemonically, utilizing combinations of letters and numbers as coded abbreviations for the text of an instruction.

The operation (OP) part specifies the particular work to be performed by central control, such as transferring data from an address in program store memory to the X register in central control. The identification constant is a digit which identifies the pertinent word in a block or group of words at a particular address in program store. This constant is not changed until the program itself is altered. (This can be accomplished only by writing new permanent magnet twistor cards as later explained.) The data address (DA) field refers to the address or location of the binary data words of the specified instruction in program store or in call store. The locations in the program and call stores are part of a joint memory list in the program instruction. The binary value of the address determines whether a specified location is in call store or in program store.

The options part of the instruction is for introducing options or required variations in the execution of an instruction. For instance, the contents of the X register in central control may be moved to an address or location in call store by the complementing (C) option. In this case, each bit of the binary word to be moved is changed from 1 to 0 or vice versa. For example, the binary word 01110110 would become 10001001. Another kind of option included in many instructions is the product masking (P) option which is illustrated in Fig. 3-17 and subsequently explained.

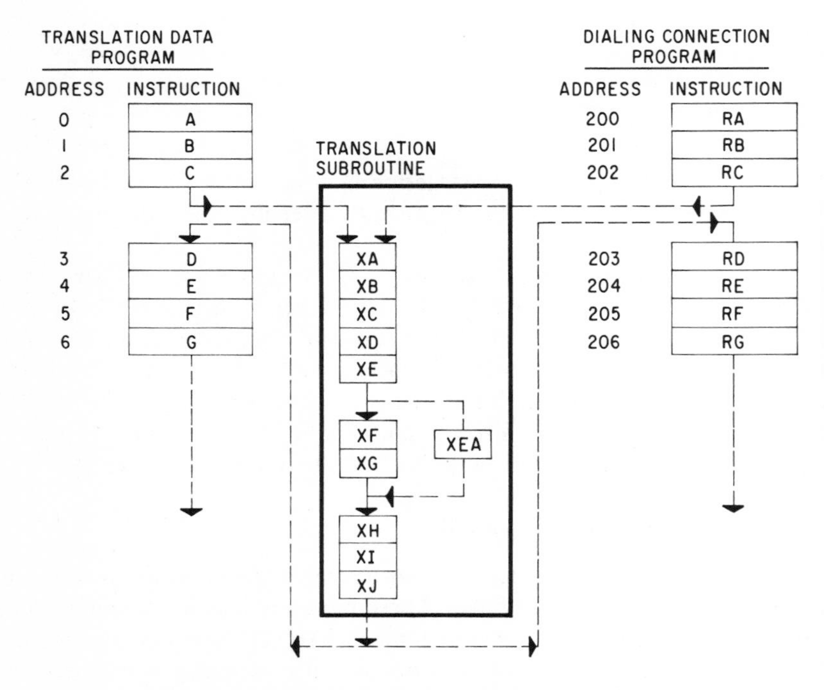


**Fig. 3-16.** Application of a Subroutine to a Program

Some segments of programs may require identical sequence of instructions at various points. To repeat the identical sequence wherever needed would involve a needless repetition and waste of program words. To preclude this situation, the particular instruction can be provided only once as a *subroutine* which is common to several programs. Figure 3-16 illustrates the operation of a translation subroutine that is common to both the translation data and dialing connection programs. It functions in the following manner: At the end of instruction C, the translation data program is transferred to the translation subroutine. The program continues through instructions XA to XJ or through instructions XA, XB, XC, XD, to XE and then through XEA to XH, XI, and XJ. From instruction XJ at the end of the subroutine, the program returns to instruction D. A similar procedure is followed for the dialing connection program in which the transfer to the translation subroutine is made from address 202 (instruction RC). In this case, the program returns to address 203 (instruction RD). Note that at the end of a subroutine, central control returns to the appropriate point in the program.

## Logic and Arithmetic Operations in Central Control

Central control processes data by manipulating arithmetic and logic quantities expressed in the binary number system which was previously explained. The arithmetic numbers contain 23 bits, designated 0 to 22, with bit 22 known as the *sign bit* because it is used to distinguish between positive and negative numbers. Addition and subtraction are the two basic arithmetic operations performed since the call processing and other system functions do not require multiplication or division. The arithmetic circuits in central control also compare, shift, and rotate binary numbers and perform certain logical operations.

The *compare operation* is related to subtraction. It involves the subtraction of two numbers to determine the sign and homogeneity of the difference as a basis of a transfer decision in a subsequent instruction. The *shifting function* is often executed to transfer two items of information, found in different positions within two data words, to a position where the two items may be logically combined. For example, the dial pulse count for a decimal digit may be accumulated in bit positions 0–3 of the K register as shown in Fig. 3-17A. However, it may have to be stored in bit positions (from right to left) 4–7, 8–11, or 12–15 according to the particular digit of the called number it represents. A shifting operation is necessary, therefore, to get the accumulated data from bit positions 0–3, to positions 4–7. Another type of shift operation is illustrated in Fig. 3-17B which shows the shifting of a 16-bit word four positions to the right. Note that each bit in this binary word is moved four positions to the right so that the underlined bits, $\underline{0110}$, in positions 0–3 are moved out of the word and deleted. Four 0 bits are added to bit positions 12–15, which were initially occupied by the first four bits of the binary word.

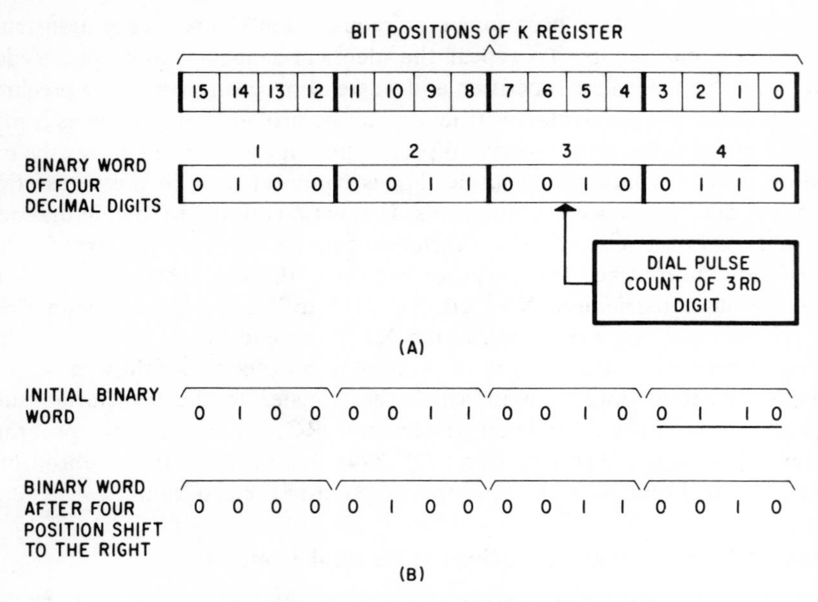


Fig. 3-17. Examples of Shift Operations

Shifts to a different number of positions, whether to the right or left, are handled in a similar manner.

*Rotation* is an arithmetic operation in which bits are passed through one end of a register in central control and reinserted at the other end but no information is lost in the process. Any number of bits or all bits in a word can be rotated as required in either direction. An example of the rotation or circular shift of binary numbers is as follows. The initial 16-bit binary word: 0100 0011 0010 0110. The binary word after being rotated four positions to the right: 0110 0100 0011 0010. (See Fig. 3-17B.) Note that the four-position rotation operation is similar to a shift to the right except that no bits are lost. Any bit that passes through the right end of the binary word will return through its left end. Similar procedures apply to rotations to the left. The rotation operation is used extensively in the network path hunt program.

## *Example of Logic Operations in Central Control*

The previous described manipulations of arithmetic and logic quantities have significant relations to the call processing functions of central control. A simplified example of such applications is shown in Fig. 3-18. This illustration refers to the processing operations involved with the state of customer lines shown in Fig. 3-11 and tabulated in Table 3-8. Recall, in this connection, that the ferrods in customer line scanning circuits are scanned at

| | BIT POSITIONS | | | | | | | | | | | | | | | |
|---|---|---|---|---|---|---|---|---|---|---|---|---|---|---|---|---|
| | 15 | 14 | 13 | 12 | 11 | 10 | 9 | 8 | 7 | 6 | 5 | 4 | 3 | 2 | 1 | 0 |
| P = PRESENT FERROD SCANNER READING | 1 | 0 | 1 | 1 | 0 | 1 | 0 | 1 | 0 | 0 | 0 | 1 | 0 | 0 | 1 | 0 |
| L = LAST SCANNER READING RECORDED IN CALL STORE | 1 | 0 | 0 | 1 | 1 | 1 | 0 | 1 | 0 | 0 | 0 | 0 | 1 | 1 | 1 | 0 |
| C = CHANGES IN STATE OF CUSTOMER LINES (EXCLUSIVE-OR LOGIC OF P AND L) | 0 | 0 | 1 | 0 | 1 | 0 | 0 | 0 | 0 | 0 | 0 | 1 | 1 | 1 | 0 | 0 |
| D = CUSTOMER LINES INITIATING CALLS | 0 | 0 | 0 | 0 | 1 | 0 | 0 | 0 | 0 | 0 | 0 | 0 | 1 | 1 | 0 | 0 |

**Fig. 3-18.** Logic Operations in Customer Line Scanning

least once every 100 milliseconds in accordance with the line scanning program.

In Fig. 3-18, the first line, P, represents the present or latest scanner readings in binary numbers received by central control from a particular group of sixteen-line ferrods. The 0 in bit position 0 (the first from the right end) indicates that the corresponding customer's line is off-hook. The 1 in the next bit position (1) signifies that its associated customer's line is on-hook. In a similar manner, the states of the other lines in the particular scanned group are denoted by the respective binary digits 0 and 1. The second line, L, symbolizes the readouts received from a call store address in which the previous scanner readings of this same group of line ferrods had been temporarily stored. Thus, the bits in line L disclose the states of these sixteen customer lines 100 milliseconds before.

The third line, C, portrays the EXCLUSIVE OR logic combination of lines P and L in accordance with the data in Table 3-8. The bits in line C that are 1 identify the particular line scanners in which a change of state has occurred between the present and last scanner readings. Additional logic circuits are utilized to ascertain which bit positions that have undergone a change of state (as indicated by line C) are presently binary 0, or off-hook. The resultant bit positions (2, 3, and 11) are indicated in Line D by a binary 1. The binary 1s in line D signify that the corresponding customer's line is now off-hook to originate a call. Central control, therefore, will proceed to connect to each originating line, in turn, a service circuit that will provide dial tone and detect the dial pulses for further processing.

## Masking and Other Data Processing Functions

An essential data processing function is to find the required binary word in a block or group of words. To accomplish this, central control sends two inputs to the index adder circuit. One input is the identifying constant of the instruction and the other is a number representing the contents of the specified index register. The index adder adds the two and the output of the adder is the address or location of the desired binary word.

In some cases, certain bits are not conveniently located in a binary data word. For instance, the binary words contained in call store are often divided into subwords, that is, a word may have several 4-bit groups, each storing a dialed digit for a particular call in progress. In such cases, masking may be utilized to isolate a subword within the word without affecting the relative bit position. The masking operation, in effect, makes 0 all but certain specified bit positions of a word. This process is illustrated in Fig. 3-19 where the bits of binary word W are converted by a mask M so that the bits in positions where mask M = 0, are changed to 0. The bits in positions where M = 1 are not changed. The resultant word is really the logical product (AND logic function) of the word W, and the mask M in accordance with the Truth Table in Table 3-3.

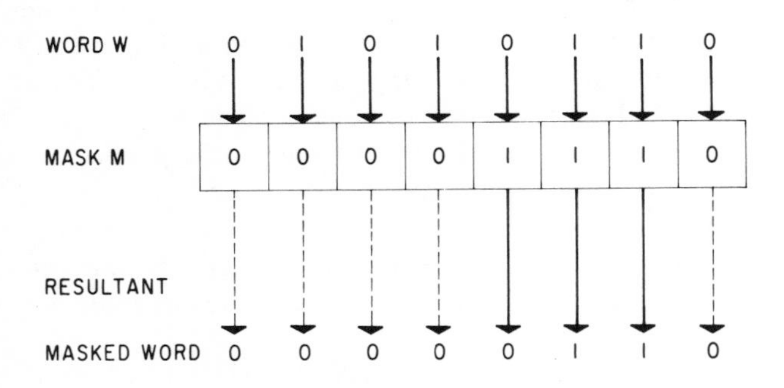


**Fig. 3-19.** Arithmetic Masking Operation of Word W and Mask M

*Masking* or the process of changing to 0 certain bit positions of a binary word, is also employed to insert information in specified bit positions of a call store address without affecting the other bit positions. This insertion masking procedure is shown in Fig. 3-20. Binary word W is inserted by mask M into address AB in call store. The procedure is that in every position where mask M has a 1, the corresponding bit from binary word W replaces the original bit of the word in location AB. In bit positions where the mask M has a 0, the corresponding bit initially in AB does not change. The resultant bits in location AB from this insert masking process are shown in the bottom row of Fig. 3-20.

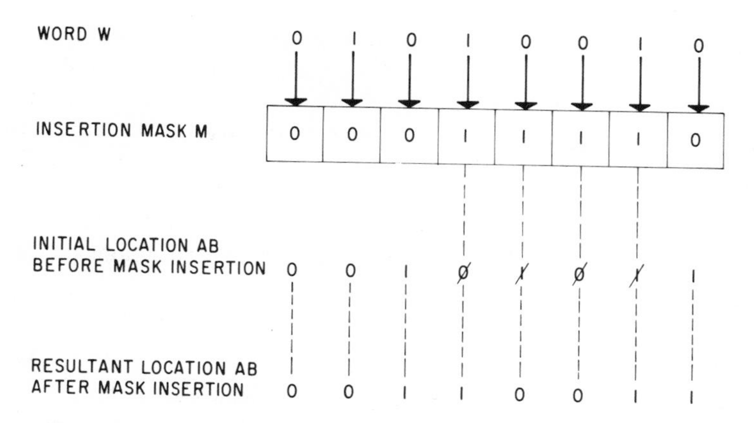


**Fig. 3-20.** Insertion Masking Procedure

Logical manipulations of binary data words follow the applicable logic functions. For instance, assume that the AND logic function per Fig. 3-3 is applied to binary words A and B. The result, as shown below, will have a binary 1 only in those bit positions where both A *and* B have a 1 (see Truth Table in Table 3-3):

| | | | | Decimal No. |
|---|---|---|---|---|
| Binary Word A | = | 0 0 1 1 0 1 1 0 | = | 54 |
| Binary Word B | = | 0 0 1 0 1 0 1 1 | = | 43 |
| Resultant AND | = | 0 0 1 0 0 0 1 0 | = | 34 |

For the logical OR function, using the same binary words A and B, the result will have a 1 in those bit positions in which A or B both have a 1 (see Table 3-4) as indicated below:

| | | | | Decimal No. |
|---|---|---|---|---|
| Binary Word A | = | 0 0 1 1 0 1 1 0 | = | 54 |
| Binary Word B | = | 0 0 1 0 1 0 1 1 | = | 43 |
| Resultant EXCLUSIVE OR | = | 0 0 1 1 1 1 1 1 | = | 63 |

The contents or binary data words of a register in central control, or a word location in either call store or program store, can be moved from one location, such as A, to also appear at another location B. The contents of location A will remain unchanged in this process. Note that data can be moved from program store but not into a program store location.

Another important data processing role of central control is the rightmost 1 function. This consists of detecting and identifying a binary 1 among a number of binary zeros in a data word. The binary 1 may signify the first customer line whose dial pulse count must be updated, or an idle trunk. For instance, if central control is searching for an idle trunk to complete a call, the binary zeros indicate busy trunks and the first or rightmost binary 1 signifies an idle trunk in the particular trunk group. In this operation, the specific binary word that stores the idle and busy states of the particular trunk group is first directed by central control to the K register. The rightmost binary 1 is then detected by an associated detector circuit which conveys the bit position of the rightmost 1 to the F register via the masked and unmasked buses as subsequently described.

## Questions

1. What are the major elements of the central processor?
2. What is meant by an address? A word? Where are they found?
3. What digits comprise the binary number system? How does the binary number system differ from the decimal system? Are other number systems possible?
4. How may numbers in the decimal system be converted to the binary form and vice versa? What are the equivalent decimal numbers for the binary numbers 0011, 1111, 101101, and 00000101?
5. How does an electrical circuit perform logic? What are the basic logic circuits? Can they be combined to form other logic circuits? Explain.
6. What particular logic circuit is extensively employed in central control? What are its main components? What is its advantage?
7. What is a flip-flop circuit? Where is it used? How are its inputs and outputs usually designated?
8. Describe the J-K flip-flop circuits with respect to the basic flip-flop. Where are they used?
9. Explain the three basic types of program instructions. Define homogeneity as used with regard to a binary word.
10. What is the value of the following binary word after it is rotated four positions to the left?

    0100 0011 0010 0110

    What is the rightmost one function?

# 4
# THE CENTRAL CONTROL UNIT

## Basic Structure of Central Control

The call processing functions of the central processor are performed by the central control unit which, as may be expected, has a very complex structure. It can be regarded as a binary digital device that performs highly complex logic operations in accordance with instructions and information received from the program store, and pertinent data from call store. The processing logic is contained within the many thousand logic circuits in central control which are designed to function at very high speeds. Central control directs the entire operations of the electronic switching system and, in effect, represents the highest degree of common control in telephone switching applications.

In order to simplify the explanations of central control's organization and operation, its principal elements are considered as consisting of *interface, control,* and *data processing* sections. Furthermore, a number of block diagrams are used to progressively present the major functions of central control with particular respect to the control and processing sections. Figure 4-1 shows the basic structure of central control including its interface units, and the major control and processing components.

All instructions and information from program store are read out or received by the buffer order word register (BOWR). This circuit serves as an interface or isolating circuit between program store and central control. The buffer order word register connects to the index adder circuit and to the control section's components. The index adder receives from the BOWR the information contained in the data-address field of the instructions transmitted by program store. The functions of the index adder will be subsequently explained. The program store address register (PAR) also interfaces with program store. It generates the addresses or locations of instructions within program store that were requested by central control through the index adder.

The data buffer register (DBR) and the index adder interface with call store. The data buffer register sends (writes) data to and obtains data (read-

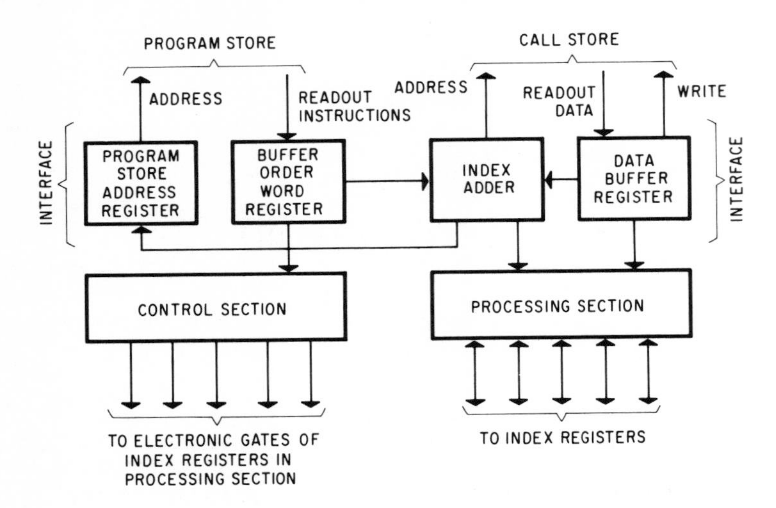


**Fig. 4-1.** Basic Structure of Central Control

outs) from call store. The data buffer register likewise connects with the index adder which, in addition to its other functions, generates addresses to call store whenever necessary to retrieve data that had been temporarily stored there by central control.

## Control and Processing Sections

The control section directs the operation of the input and output electronic gates that are associated with the various registers in the processing section of central control. The block diagrams in Fig. 4-2 present further details of the processing and control operations. Note that input and output gates are shown with the various index registers. In subsequent diagrams, these gates will be omitted for simplicity purposes because the symbol for a register will include these electronic gates.

The order word register (OWR) with the order word decoder (OWD) are used to control, in part, the execution of instructions received by the buffer order word register (BOWR) from program store. These circuits are provided because of the overlap between the basic 5.5 microseconds time cycle of program store's operations, and the approximately 10 microseconds required for processing by central control. This time interval includes the period from the receipt of instructions from program store until the specified data from call store is read out into the data buffer register and processed by the processing section. As a result, there will be a time overlap in the execution of two consecutive instructions. The buffer order word decoder (BOWD) and the order word decoder, consequently, are utilized to simultaneously control the execution of the two consecutive instructions.

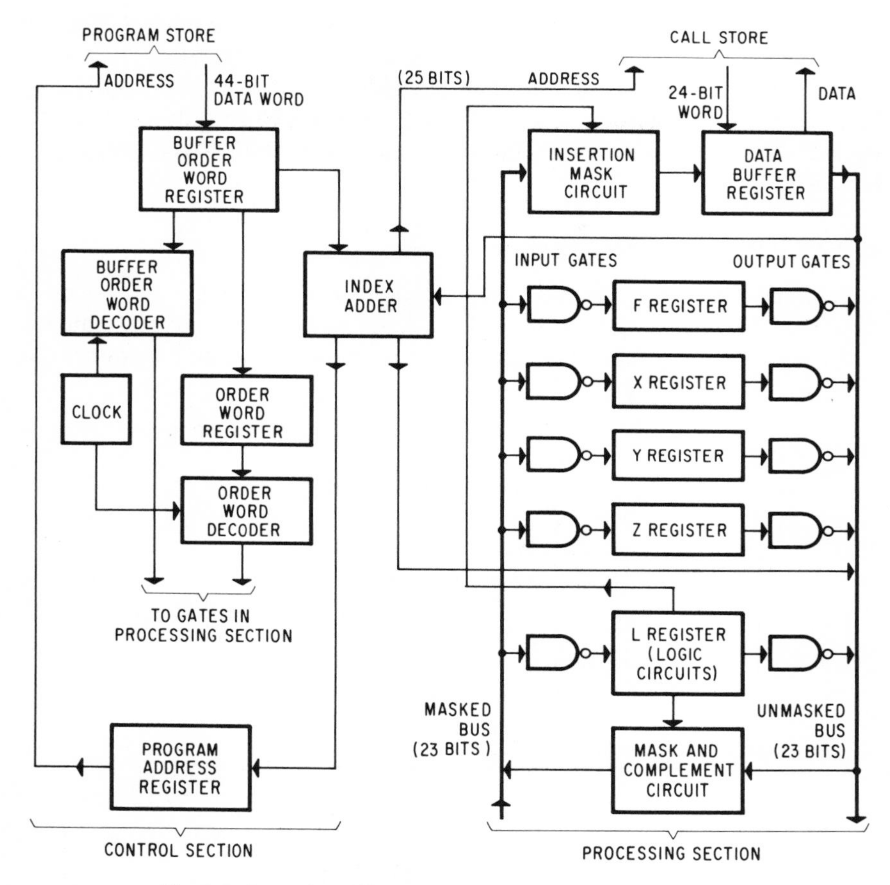


**Fig. 4-2.** Control and Processing Sections of Central Control

In addition, the buffer order word decoder (BOWD) and the order word decoder control the flow of processing information within central control by operating the input and output electronic gates associated with the registers in the processing section as illustrated in Fig. 4-2. This electronic control function is termed *gating,* and it will be frequently used in this book. For instance, data from call store first is gated to the data buffer register (DBR) under control of the order word decoder. The data is next conveyed over the unmasked bus to the specified register circuit, for example the X register, because its input gate was operated by the order word decoder. In a similar manner, the order word decoder would operate the output gate of the Y register so that its recorded data may be sent to call store for temporary storage.

*The aforementioned operations of the control section within central control may be summarized as follows:* The BOWD controls the addressing of call store. The order word decoder, on a readout instruction, controls the gating of information from call store to the data buffer register and, via the unmasked and masked buses, to the specified index register in the processing section. The order word decoder, on a written instruction, controls the gating of data from an index register and then via the buses to the data buffer register and call store.

## Processing Binary Data Words

Each register in the processing section of central control can store a binary data word of up to 23 bits. The binary word may have originated in call store or in part of the instructions that were received from program store.

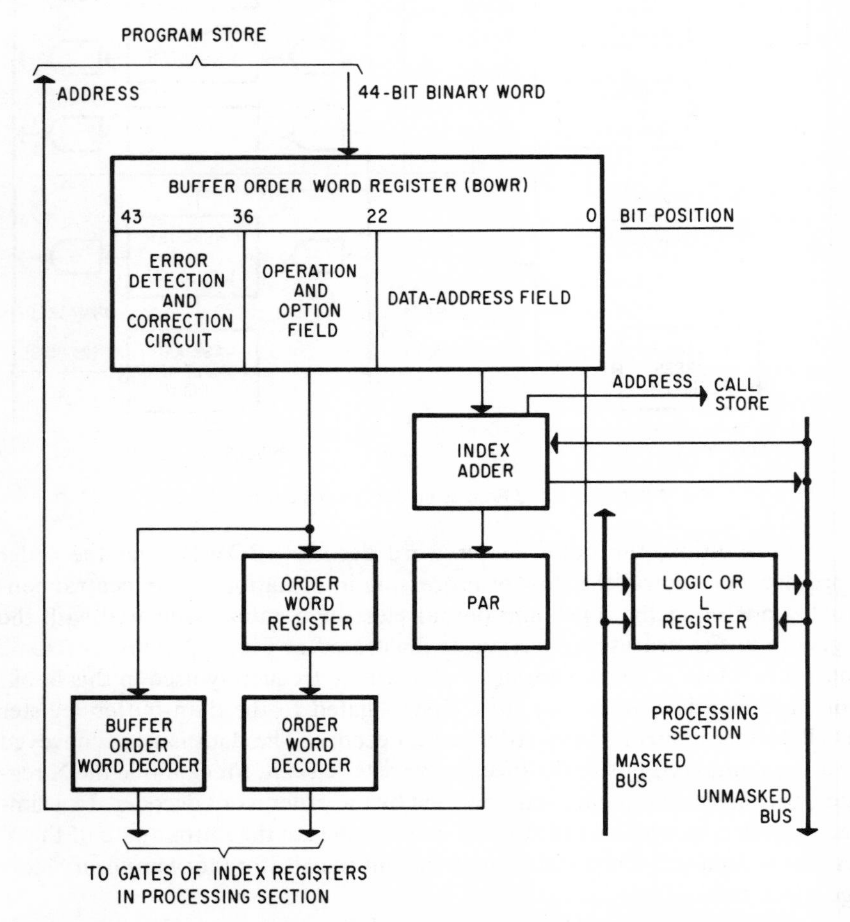


**Fig. 4-3.** Division by Central Control of a 44-Bit Binary Word Readout from Program Store

Binary words read out from call store actually consist of 24 bits which is the capacity of the data buffer register (DBR). The extra bit is used for parity check purposes and remains within the data buffer register.

Binary words from program store contain 44 bits as received by the buffer order word register (BOWR). These 44 bits are divided into three groups which are called the data-address field, operation and option field, and error detection and correction circuit as shown in Fig. 4-3. The data-address field, depending upon the type of instruction, may contain bits 0 to 22. Bits 23 to 36, inclusive, normally make up the operation and option field.

The remaining bits, 37 to 43, are used by the error detection and correction circuit. The binary data words read out from program store are encoded (using the Hamming code) so that single errors can be detected and corrected. If a double error should be discovered, the particular instructions from program store will be reread. Likewise an error in an address will cause the instructions to be reread.

Referring to Fig. 4-3, bits 0 to 22 in the data-address field are gated, under the control of timing circuits and in accordance with the program instructions, to either the L register or to the index adder for further processing. The index adder circuit may modify, when required, the data-address field by adding to it the contents of a specified index register. The logic or L register is connected to the unmasked bus and the masked bus so that this register also may be controlled and read similar to any index register.

The 14 bits, 23 to 36, inclusive, in the operation and option field, are dispatched to the order word decoder through the order word register. These particular bits also are sent to the buffer order word decoder (BOWD) as indicated in Fig. 4-3. These 14 bits in the operation and option field serve to control the gating of the several index registers in the processing section, which are illustrated in Fig. 4-2.

## Masking Functions

The 23-bit word length of an index register in the processing section of central control is longer than most of the data quantities handled with call store. For instance, a single binary-coded decimal digit, such as 8, is 4 bits long. Therefore, up to five of these data quantities may be packed into one binary data word. In order to efficiently handle such partial data words, masking facilities are provided in central control for use in accordance with program instructions.

Masking functions are provided by the mask and complement circuit, the insertion mask circuit, and the masked and unmasked buses which are shown in Fig. 4-2. Most of the masking operations are accomplished within the mask and complement circuit. The *mask circuit* has two inputs: the L or logic register's output which controls the masking function, and the unmasked bus which connects to the gates at the outputs of the various index registers. The output of the mask and complement circuit is the masked bus which connects to the input gates of the several index registers. The *complement circuit* is in series with the mask circuit. It is utilized to change each bit in a data binary word from 0 to 1, or from 1 to 0 whenever the contents of an index register are moved from central control to call store for temporary memory storage. Some of the masking operations are discussed in Chapter 3.

*Insertion masking,* provided by the insertion mask circuit, is another form of masking that is frequently used. It allows all but a selected group of bits in a specified register to remain intact. The insertion mask circuit is associated with the data buffer register (DBR), as indicated in Fig. 4-2, because insertion masking is often employed when only a portion of a binary word in call store is to be modified. The L register also controls the insertion mask circuit because, in most cases, the bits to be inserted and their associated positions have been recorded in the L register in connection with a previous masking operation.

It should be noted that call store has no circuits for performing insertion or other masking operations. Consequently, masking is entirely a function of central control that is accomplished in two steps. First, the binary word requiring the insertion of information is read from call store. Next, the required information is inserted in the word and the completed binary word is sent back to call store. For example, assume that an instruction from call store specified that certain data be read out from call store to the X register. The requested binary data word would enter the data buffer register (DBR) from call store. It would be gated over the unmasked bus to the mask and complement circuit and then onto the masked bus to the X register. Observe that the data buffer register would contain the data word exactly as received (read out) from call store. The binary word received in the X register, however, may have been modified by the mask and complement circuit if so required by the program instructions. The modified word in the X register, subsequently, may be sent back over the unmasked bus, through the mask and complement circuit, the masked bus, the insertion mask circuit, and then through the data buffer register (DBR) to call store. In this case, the modified binary data word could be further altered as it passes through the insertion mask circuit, depending upon the program instructions.

## Program Instruction Rudiments

For simplicity, it has been assumed that all instructions originate in program store and that data is stored in call store. During operations, considerable data, especially translation information pertaining to central office codes, are stored in program store. Moreover, the same type of instruction to central control is used to obtain readouts of data from program store and call store. To read (interrogate) either program store or call store for data is a complicated process, particularly since overlap operation of instructions is in-

**Table 4-1.** Program Instruction Format

| Operation | Program Identification Constant | Index Register Address | Masking Option |
|---|---|---|---|
| MX | 4 | Y | |
| MY | 2 | X | PL |

volved. It is not within the scope of this book to attempt to explain all of the procedures involved. The necessary explanations will thus be presented in a series of simplified steps covering the essential program elements.

Program instructions for electronic switching systems, such as the No. 1 ESS, may be divided into three main categories designated operation, program constant, and index register. In addition, a masking option is included in many instructions. Table 4-1 illustrates this basic composition of a program instructions.

An important operation associated with the program instruction is the indexing process performed by the several index registers. These registers are the X, Y, Z, F, J, K, and the data buffer register. Indexing is one method of obtaining an address or location in the call store memory by adding the program constant in a particular program instruction to the binary information previously recorded in one of the index registers.

The indexing process is also used on connection with a list of binary words that may be recorded in consecutive call store addresses or locations. For instance, the program instruction would specify the address of the first binary word in the list. Then, the contents of a designated index register would indicate the position of the desired binary word in this list with respect to the first word. In this manner, the same program instruction may be used a number of times to affect different binary words of a list, thereby reducing the number of required program instructions.

## Example of Program Instruction Operations

Central control can go to the index registers for information at any step in a program instruction. For example, the Y register is assumed to contain the address of a block of binary words in call store that are utilized for accumulating the dialed digits of a call in progress. It is required that the reading of the fourth word in the particular block of words be transferred to the X register for further use in the processing sequence. The program instruction for this purpose may be summarized as: Read out the data from the fourth word of a block of binary words in call store whose starting address is now in the Y register, and store this reading (data) in the X register.

The actual instructions to central control from program store are in the form of letter and figure symbols or mnemonics. For the above example, the instructions from program store would be written, "MX, 4, Y" as indicated in Table 4-1. The mnemonics, MX, mean memory to X index register. In this respect, the program instructions which specify a read (interrogate data) or write (insert data) requirement in call store memory, contain the letter M in their mnemonic representation, and are designated M instructions. The operating procedure for the aforementioned example, therefore, may be described as follows with reference to Table 4-1 where they are shown under the respective parts of the program instructions:

1. *Operation* = MX (call store memory to X index register). Take data from a specified location in call store memory and store it in the X index register.
2. *Constant* = 4 (binary word with required data). The required data is in the fourth word of a block of binary words in call store whose address is in the index register address data.
3. *Index Register Address* = Y (address information). The starting address, or location in call store of the specified block of binary data words containing the required data, is in the Y register.

## Example of Masking Instructions and Operations

To illustrate the application of masking in program instructions, let us assume that central control is directed to ascertain whether or not the first digit dialed was zero (call to the operator). The particular program instructions would be written "MY, 2, X, PL" in order to perform the following tasks:

1. *Operation* = MY. Take data from a specified location in call store memory and store it in the Y index register.
2. *Constant* = 2. The required data is contained in the second word of a block of binary data words in call store at the address in the index register address.
3. *Index Register Address* = X. The starting address or location in call store memory of the specified block of binary data words, containing the word with the desired data, is in the X register.
4. *Masking Option* = PL (Product and logic masking). Mask out all bits in the secondary binary word (constant 2), except the four bits used for the first dialed digit, and store the resultant word in the Y index register. Compare this binary word, using the L (logic) register with the decimal value of 10 (binary 1010). This value is the number of pulses counted when the digit 0 is dialed.
5. *L Register Logic Operations.* If the two compared numerical quantities are equal (first digit dialed is zero), central control will transfer to another type of program at the particular address specified in the instructions. If the two compared quantities are not equal, cen-

tral control will continue with the existing program sequence because the first digit dialed was other than zero.

The mask and complement circuit and the insertion mask circuit are used to perform the masking options that are specified in the program instructions. In addition to the PL option described in the above example, a PS option is provided by the data-address field of the program instruction. With the PS option, the information in the data-address field is gated directly to the L register, as shown in Fig. 4-3, instead of entering the index adder. The insertion masking and complement options are employed in connection with the transfer of information from a register in central control to an address or location in call store. Two types of insertion masking are normally provided. They are written as EL and ES in the program instructions. Complement masking may be defined as storing a negative number as the complement of the positive number of the same value. It is written as PLC in the program instructions.

## Program Instruction Transfers

Addressing or locating instructions in program store is a function of the program store address register (PAR) as previously mentioned. Instructions from program store normally advance in sequence from one particular instruction to the next. This process is controlled in central control by the increment circuit which is associated with the program address register (PAR). The increment circuit is comprised of the add-one register (AOR) and the add-one logic (AOL) circuit as shown in Fig. 4-4. Some functions of the increment circuit are illustrated in the following example.

Let us assume that a certain program instruction, being processed in central control, came from address 206 in program store. In order to receive the next instruction at address 207, it will be necessary to add one (1) to address 206. This add-one function or incrementing operation is performed by the add-one register (AOR) and the add-one logic (AOL) component of the increment circuit associated with the program address register (PAR). This incrementing or adding of 1 continues until the completion of the various program instructions involved unless a transfer instruction is received by central control.

The increment circuit also may be utilized with index registers, such as X, Y, and Z, as an option of the program instructions. The incrementing option is often employed when the same processing task is being performed on a number of successive addresses or memory locations. The specific index register is set to the value of the first memory address and the index register is incremented by 1 as successive words are read out from call store. The input and output of the increment circuit, for this function, connect to the unmasked bus and the masked bus, respectively, as shown in Fig. 4-4. There-

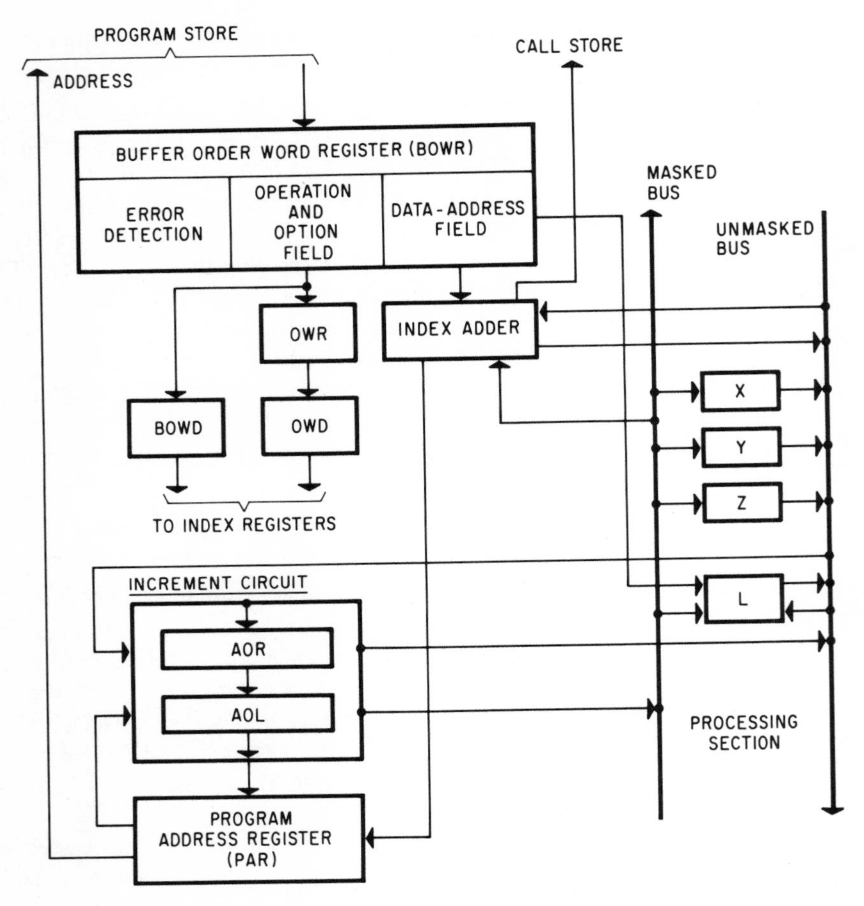


**Fig. 4-4.** Use of Increment Circuit in Central Control

fore, any index register may be gated to the increment circuit as may be specified in the program instruction.

The increment circuit can be restrained or inhibited by a transfer instruction contained in the program instructions received by central control. A transfer instruction is an instruction from program store directing central control to go to another set of instructions instead of to the immediately following instruction. In such case, the address of the new instruction would be derived from the output of the index adder because it is the place where the contents of the program instructions are combined with the contents of index registers and, consequently, also with memory readings. The index adder connects to the program address register (PAR) which will generate the new address in program store for this direct transfer of instructions. These interconnections are depicted in Fig. 4-4.

Call store, although normally the source of data for central control, also can be employed as a source of instructions. For instance, a transfer instruction from program store may specify an address in call store for the next instruction. This procedure, designated an indirect transfer, usually is denoted by an M suffix in the index register address part of a program instruction, as shown in Table 4-1. In this event, the particular instruction from call store would be transmitted to the buffer order word register (BOWR) instead of to the data buffer register which interfaces with call store. The BOWR normally interfaces only with program store. In addition, a sequencing circuit is used to assemble the information received in the buffer order word register (BOWR) from call store because, as later explained, additional time cycles (total of four) are needed by central control to execute program instructions received from call store.

## Sequencing Circuits

A number of sequencing circuits or sequencers are associated with the decoders, index registers, and other pertinent elements of central control as portrayed by the block diagrams in Fig. 4-5. A sequencing circuit generally performs at least two of the following functions:

1. To inhibit or restrict the operation of decoders.
2. To increase the number of time cycles required for the processing of an instruction from call store or data from program store.
3. To generate the necessary gating signals during the period of the additional time cycles when the decoders are inhibited.

Some of these functions of sequencing circuits are illustrated in the following example.

Assume that an instruction recorded in the order word register (OWR) specifies that a transfer is to be made to another set of program instructions. It will be necessary, therefore, to immediately restrict the indexing operations in the buffer order word register (BOWR) because the present instruction in the BOWR is not going to be used. Also, an additional time cycle will be needed for transmitting the instructions contained in the specified transfer address in program store to the buffer order word register (BOWR). To accomplish these actions, a sequencing circuit, also termed a sequencer, is used to restrict or inhibit the buffer order word decoder (BOWD) and the order word decoder (OWD), and to generate the necessary gating signals during the time cycles when these decoders are inhibited.

A sequencing circuit likewise is required when translation information if read out from program store because this type of information is treated as data and not as program instruction. Thus, a sequencing circuit is needed to provide the three time cycles required for this type of processing. Additional sequencing circuits, as later explained, are provided to perform a number of other functions including the following:

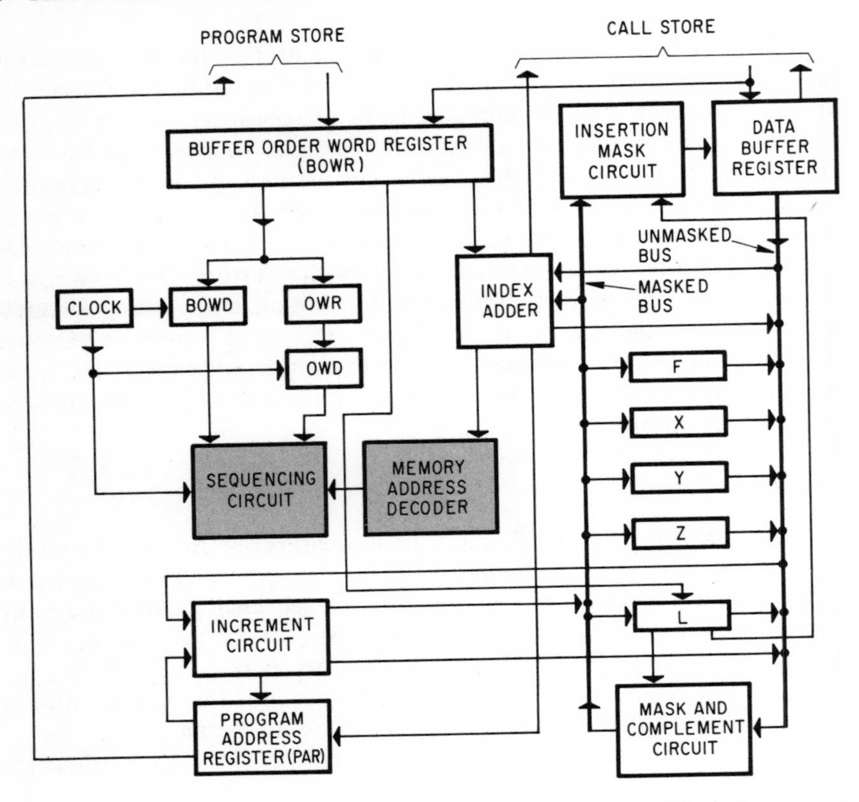


**Fig. 4-5.** Use of Sequencing Circuit with Decoders in Central Control

1. Directing program interrupt signals due to trouble conditions or the need to transfer to another processing task, for example, scanning for dial pulses.
2. Repeating the readout of a binary word from program store because of errors that cannot be corrected.
3. Repeating the readout of a binary word from call store because of indicated trouble.
4. Controlling communications with peripheral units.
5. Mutating readouts from a call store that is associated with a signal processor.
6. Stopping and then restarting central control.

## Data Instructions from Program Store

In addition to program instructions, considerable translation information or data concerning the routing of calls is stored in program store. Addresses or locations of memory lists (binary words containing instructions or data), as previously mentioned, are part of a joint memory list with certain series of addresses allocated to call store and others to program store. The memory address decoder is utilized to ascertain whether the address recorded in the index adder circuit pertains to call store or to program store. If the address relates to call store, the associated instruction is executed by central control within one time cycle in the normal manner. However, if the address is in program store, the memory address decoder functions to activate a special sequencing circuit. These additional components of central control are shown in Fig. 4-5, which also includes block diagrams of the other components that have been discussed.

The sequencing circuit, in the case where the address is in program store, serves two main purposes. It prevents the translation data, sent by program store from being incorrectly considered as program instruction, and it provides the necessary three time cycles for handling this data within central control. The essential processing steps involved are outlined in Table 4-2 and also explained in the following example.

Let us assume that during the initial time cycle a new program instruction E is contained within the order word register (OWR), that its address in program store is 201, and that instruction E requires that certain translation data, including binary word W, be read out from program store. The index adder circuit, therefore, sends address 201 to the program store address register (PAR) for transmission to program store. At the same time, the memory address decoder recognizes the address 201 as being in program store and activates a sequencing circuit to inhibit the buffer order word decoder (BOWD). As a result, the existing instruction A in the buffer order word register (BOWR) is disregarded. These actions are shown as cycle period 1 in Table 4-2.

**Table 4-2.** Time Sequence Steps in Processing of Binary Word Translation Data from Program Store

| Binary Word from Program Store | | | Program Store Address |
|---|---|---|---|
| Cycle Period | In BOWR | In OWR | In PAR |
| 1 | A* | E† | 201 |
| 2 | W | E | 321 |
| 3 | A+1‡ | E | 322 |
| 4 | A+2§ | A+1 | 323 |

* Initial program instruction at address 321 in program store which was inhibited by action of sequencing circuit because of instruction E.

† Program instruction requesting readout of binary data word W from program store at address 201.

‡ Program store instructions at address 321.

§ Program store instructions at address 322.

In the second time cycle, the requested translation data (binary word W) is received from program store by the buffer order word register (BOWR) and immediately sent via the index adder and the unmasked bus to the data buffer register which interfaces with call store. At the same time, program store address 321 of program instruction A, which was kept in the add-one register of the increment circuit, is again sent to program store by the program address register (PAR). Note, as tabulated in Table 4-2, that program instruction E continues to remain in the order word register (OWR) during the first, second, and third cycle periods.

During the third cycle period, program instruction E, which had been contained within the order word register (OWR), is executed just as if the translation data (binary word W which was previously placed in the data buffer register) had been read out from call store instead of from program store. Simultaneously, the next program instruction, A+1, is received in the buffer order word register (BOWR) from program address 321 and dispatched to the index adder for necessary indexing operations. At the same instant, the address of the succeeding program instruction, 322 (321+1), is placed into the program address register (PAR).

Normal program instruction operations are resumed in the fourth cycle period, as indicated in Table 4-2, with the receipt of program instruction A+2 in the buffer order word register (BOWR). In the same period, program instruction A+1, which is in the order word register (OWR), will be executed, and the address of the next program instruction, 323 (322+1), is inserted by the increment circuit into the program address register (PAR) for transmission to program store.

## Call Processing Time Cycle and Overlap Operations

The actual circuit actions for each step in the processing of a call take place in central control which, as previously mentioned, is a synchronous machine capable of performing only one action at a time. These actions are made in a cycle period of 5.5 microseconds. This time cycle is established by the electronic clock depicted in Fig. 4-5. This clock governs the operations of the BOWD and OWD decoders, sequencing circuits, related index registers, and other components. In effect, the clock establishes the gating period for the operation of the various decoders and index registers.

During the 5.5-microsecond period, one of the many thousands of binary words that make up the instructions in program store directs central control in a particular call processing task, or to carry out some automatic maintenance procedure. These different operations follow a precise schedule because central control can execute only one instruction at a time, and each step may require a different time duration. For example, customer lines are scanned periodically to detect the dialing of digits. This scanning must be completed fast enough to ensure that no dial pulses are missed. The switching network, on the other hand, operates at a slower rate so that instructions to make connections in the switching network are issued at a correspondingly lesser interval.

In most cases, a 10-microsecond cycle would be required for each step because of transmission delays in cables, response time of the memory devices in program store and in call store, and other factors as tabulated in Table 4-3. This 10-microsecond period is based on the processing of one instruction at a time by central control (from the preliminary to the execution stages) before another instruction is received from program store. In order to maintain the desired 5.5 microsecond time cycle, central control in the No. 1 ESS is designed to operate in the overlap mode. For this purpose, an instruction is divided into two parts, called the preliminary and execution stages, and central control processes two instructions simultaneously as described below and diagramed in Fig. 4-6.

When a particular program instruction, D, arrives at the buffer order word register (BOWR), the program address register (PAR) sends the address of the following instruction, D+1, to program store, and the preliminary stage of incumbent instruction, D, is processed. This overlap operation in central control may be better understood if it is assumed that program instruction D is located at address 251 in program store, and that the format of instruction D is "MX2Z." Therefore, "2Z" is the preliminary stage and "MX" is the execution stage of instruction D.

During the preliminary stage of program instruction D, the unmasked bus is used for the indexing operations and any necessary index register modi-

**Table 4-3.** Time Period for Call Processing Operations in Central Control without Overlap Mode

| Call Processing Operation | Approximate Time in Microseconds |
|---|---|
| Resetting of instruction register in program store | 0.50 |
| Transmissions of instructions from program store to BOWR and the stabilization of its output | 2.25 |
| Decoding operations | 0.50 |
| Gating of index registers into the index adder and sending the address to call store | 1.50 |
| Maximum readout time from call store into data buffer register (B) | 3.75 |
| Gating of call store readout from the data buffer register (B) to the specified index register (such as Z) | 1.50 |
| Total Time of Operations | 10.00 |

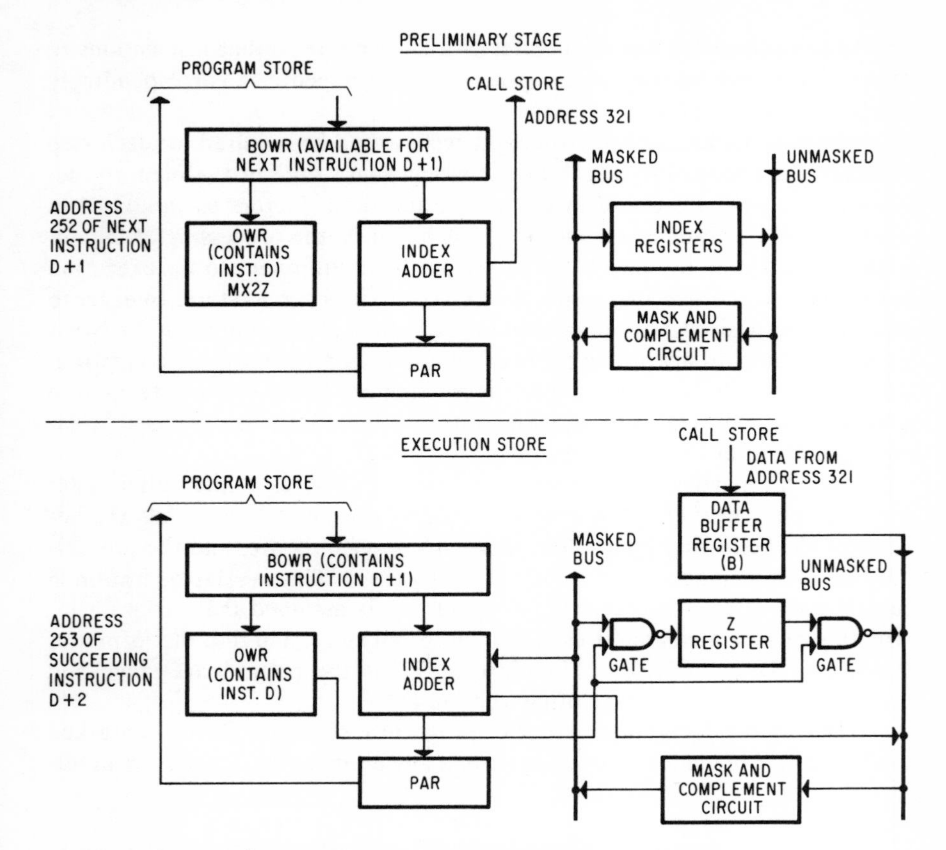


**Fig. 4-6.** Processing in Preliminary and Execution Stages of Program Instructions

fications are sent over the masked bus to the index adder. These operations are controlled by the buffer order word register (BOWR). At the same time, the following tasks are performed within central control:

1. The index adder develops the address, for example, 321, of the required data in call store and sends this address to call store with a readout request.
2. The program address register (PAR) sends the address 252 (251 +1) of the next instruction (D+1) to program store.
3. Program instruction D (comprising format MX2Z in this example) is transferred from the buffer order word register (BOWR) to the order word register (OWR). Therefore, BOWR becomes available to receive the next program instruction, D+1, located at address 252 in program store.

In the following execution stage, the data from address 321 in call store is received by the data buffer register (DBR) and gated to the Z index register, via the unmasked bus, the mask and complement circuit, and over the masked bus under control of the order word register (OWR). At the same instant, the preliminary stage operations of checking, indexing, and index register modifications for the next program instruction, D+1, that was received from address 252 in program store are performed. Likewise, address 253 of succeeding instruction D+2 is sent to program store by the program address register (PAR). During the overlap time cycle when the masked and unmasked buses are jointly used by the two instructions being processed in central control, the preliminary stage of the next instruction uses only the unmasked bus while the existing instruction in the execution stage uses the masked bus. This arrangement is employed to prevent mutual interference.

## K Register and Accumulator System

To perform the many additions and logical combinations of two quantities that are required in central control, the K accumulator register and associated accumulator circuitry are provided. This accumulator system, illustrated in Fig. 4-7, consists of the K addend and K augend registers, and the logic and arithmetic circuit, in addition to the K register. Also, a rotate and shift circuit and a detect first-one circuit are associated with the K register. The functions of these circuits will be explained later.

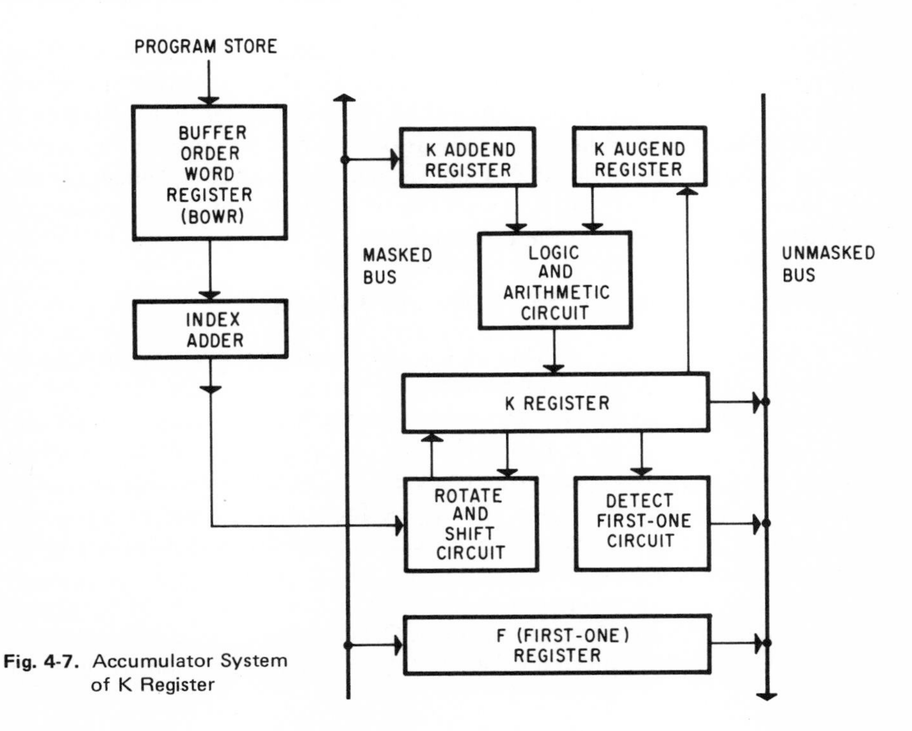


**Fig. 4-7.** Accumulator System of K Register

The K register, in addition to its use as an index register like the X, Y, and Z registers, also controls various logical and arithmetic operations. For instance, its binary contents can be gated into the K augend register and then combined with the contents of the K addend register in the logic and arithmetic circuit. This combining action normally would employ simple addition or logic functions such as AND, OR, or EXCLUSIVE OR that were described in Chapter 3.

The binary contents of the K addend register are received, via the masked bus, from the data-address field of the program instruction that had been deposited in the buffer order word register. This arrangement is depicted in Fig. 4-3. Moreover, when several binary numbers are to be added together, this must be accomplished by adding one at a time. During this process, the partial sums are accumulated in the K register until the final sum is attained.

Binary data in the K register also can be rotated and shifted by the rotate and shift circuit. The shifting action is required to line up two items of information, found in different positions within two binary data words, to a position where the two items may be combined logically. The rotation operation is similar to the shift operation except that, for a left rotate, the contents of the most significant bit are shifted back into the least significant bit instead of being shifted out, and vice versa for a rotation to the right. These rotation and shifting operations are further explained in the section "Logic and Arithmetic Operations in Central Control" in Chapter 3.

The detect first-one circuit is employed to detect and identify a 1 in the midst of a group of zeros in a binary data word in the K register. In effect, it determines the position of the rightmost bit in the K register that is equal to 1. The binary number that identifies the position of this bit is then put into the F or first-one register, via the unmasked bus, the mask and complement circuit, and the masked bus. A circuit to reset this bit in the K register will receive its selection information from the F register.

## Functions of CS and CH Flip-Flops and J Register

A pair of flip-flops, designated CS and CH, also connect to the masked bus and to the logic and arithmetic circuit of the K register, as shown in Fig. 4-8. Flip-flops are solid-state devices comprising a combination of logic circuits that provide for the temporary storage of binary digits or bits of information as explained in Chapter 3. The CS and CH flip-flops normally function when a compare operation is being performed in the accumulator system. In this type of operation, the difference of two binary numbers to be compared is computed by the logic and arithmetic circuit.

The result of the computation, however, is not gated into the K register but is utilized only to control the CS and CH flip-flops. These flip-flops also may be controlled during the execution stage of an instruction whenever a

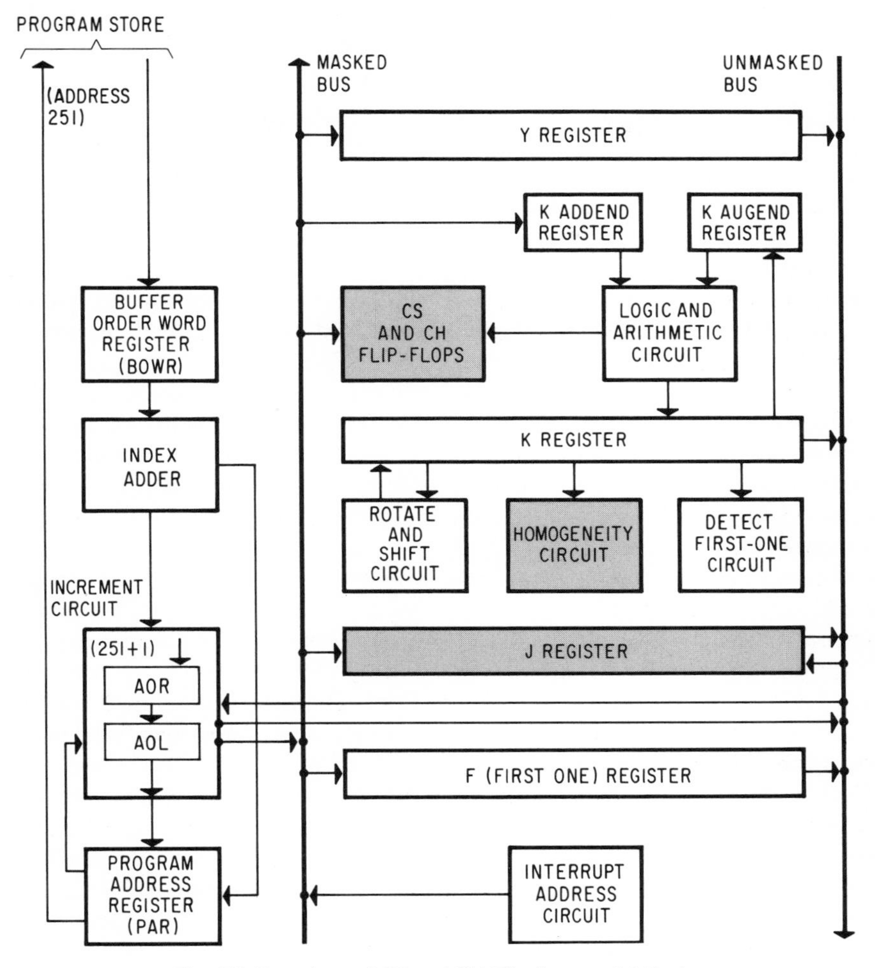


**Fig. 4-8.** Functions of CS and CH Flip-flops and J Register

binary data word is gated, via the masked bus, to a register other than the K register. In this case, the sign and homogeneity of the binary word will control the CS and CH flip-flops.

These flip-flop and K register functions are related mainly to conditional transfer instructions received from program store. This type of program instruction usually directs central control to proceed to another set of instructions, instead of to the immediately following instruction, if a certain data word or bit of data having a predetermined value (such as arithmetic zero) has been deposited in the K register or in the CS and CH flip-flops. If the binary word or bit does not have that value, the transfer is not made and

**Table 4-4.** Format for Conditional Transfer Instruction

| Operation | Program Identification Constant | Index Register Address | Masking and Other Options |
|---|---|---|---|
| CW | 10 | Y | |
| TCA | 251 | | J |

the next following instruction is executed in the normal manner. Arithmetic zero, the usual predetermined value, includes +0 bits (all zeros) and −0 bits (all ones). In both cases, all bits are identical and, therefore, the binary word has a homogeneity of 1.

An example of an conditional transfer instruction format utilizing the CS and CH flip-flops is outlined in Table 4-4. The instruction format used for this example is similar to that shown in Table 4-1. Referring to Table 4-4, the program instruction, "CW, 10, Y" tells central control to compare the program identification constant 10 with the contents of the Y register, and to gate the result to the CS and CH flip-flops. These flip-flops are employed for this comparison operation because the K register is not used for this particular program instruction.

Central control performs the comparison operation by subtracting 10 from the bit contents of the Y register, placing the result on the masked bus, and gating it only to the CS and CH flip-flops. The condition transfer instruction, "TCA, 251, J," which follows is then executed. It directs central control to transfer to the program instruction located at address 251 in program store, if the result of the previous comparison action by the CS and CH flip-flops indicated arithmetic zero. In other words, if the contents of the Y register equaled the program identification constant 10, arithmetic zero would result. Otherwise, central control would proceed to execute the next instruction which would be at the address in the program address register (PAR).

The J option indicated in Table 4-4 is the return address or jump option often associated with transfer instructions. Most transfer instructions frequently occur in the middle of a program. They usually involve a transfer to a subroutine in connection with performing common tasks as explained in Chapter 3. In order to direct the subroutine where to return control to the original program instruction, the J register is provided as shown in Fig. 4-8. When a conditional transfer, as in the foregoing example, is executed, the add-one register (AOR) in the increment circuit contains the contents of the next program store address (251 + 1) to be sent to the program address register (PAR). This information, therefore, will be gated to the J register via the unmasked bus. Consequently, when the conditional transfer takes place, the address of the next program instruction will be saved in the J register. Block diagrams of the various components discussed herein are shown in Fig. 4-8 together with the related elements of central control.

## Buffer Bus Registers

In addition to the different index registers and the CS and CH flip-flops previously described, a number of other flip-flops are provided in central control. These flip-flops are arranged in 22 groups, called buffer bus registers, which are associated with the data buffer register. Figure 4-9 shows the arrangement of the buffer bus registers in central control.

The buffer bus registers or flip-flop groups are utilized for maintenance purposes and with match circuits as later explained. Each flip-flop group is assigned a distinct address. Thus, when this address is generated by an instruction from program store, the memory address decoder (illustrated in Fig. 4-5) will operate the gates to and from the particular flip-flop group or buffer bus register. In this way, the control of the buffer bus registers does not require additional space in the program instruction format.

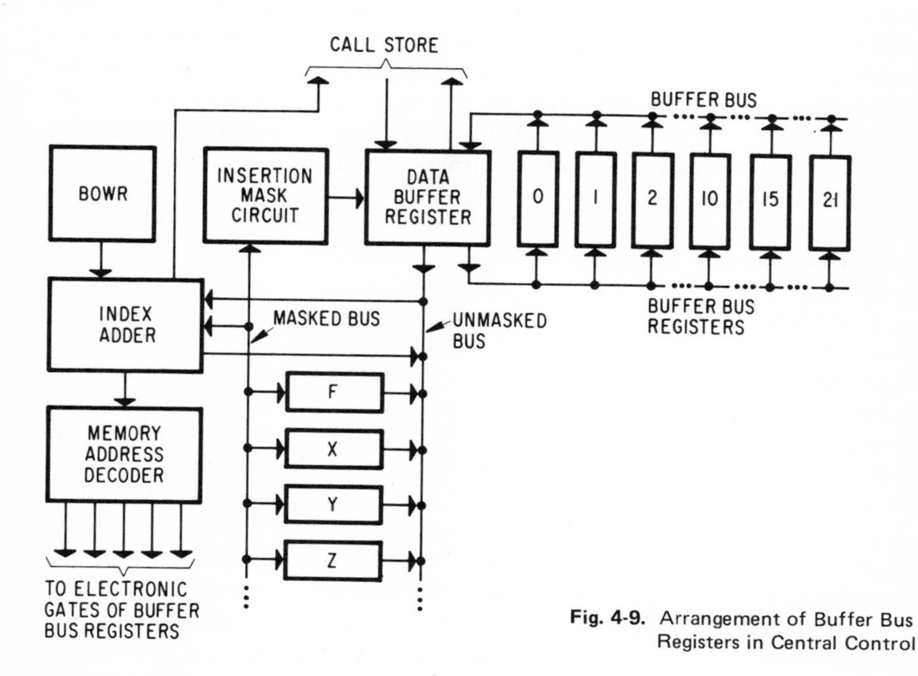


**Fig. 4-9.** Arrangement of Buffer Bus Registers in Central Control

## Interrupt Signals

Interrupt signals may be generated within central control to interrupt a program instruction being executed, and to initiate a transfer to another program instruction associated with the interrupt signal. The interrupt signal is normally originated by the following two conditions:

1. Detection of a trouble condition by the maintenance circuits. This necessitates an immediate interruption in the program being exe-

cuted in order that the trouble can be analyzed and the required corrective measures be taken before the trouble condition causes errors in call processing operations.

2. Reception of the 5-millisecond clock pulse which initiates the dial pulse scanning program and similar functions that must be performed every 5 to 10 milliseconds.

The interrupt signal causes the interrupt address circuit, which is connected to the masked bus as illustrated in Fig. 4-8, to generate the particular program transfer address required for the type of interrupt signal that was received. A total of ten program addresses are provided for this purpose in the No. 1 ESS. The transfer to the specified address is accomplished under the control of a special sequencing circuit.

The call processing operations in progress within central control are not disrupted when an interrupt signal is received, it is merely postponed. Moreover, the program transfer initiated by the interrupt signal takes place only after the completion of the instruction in progress at the time that the interrupt signal was received. These actions are possible because all pertinent information in the various registers within central control is stored in call store and, therefore, can be retrieved when the interrupted program is resumed. The transfer back to the program instruction that had been interrupted is controlled by a sequencing circuit.

## Matching Circuits

In most operations, two central control units, designated 0 and 1, execute identical tasks within the same program sequences. One central control unit is assigned as the active unit and the other is the standby equipment. The individual data processing steps for each task are closely synchronized because the microsecond clock in the duplicate central control unit is driven by only one of the two 2 MHz crystal oscillators. The 2 MHz crystal oscillator in the active central control unit governs the microsecond clocks in both units.

Both central controls start functioning by placing the same address in the program address register (PAR) to simultaneously obtain and execute the related instructions from program store. Each central control continues the call processing tasks in step with the other. For instance, the same data are read from the call store memory and from the line, junctor, and trunk scanners. Also, identical data processing steps are performed on the received data from call store and each central control unit makes similar decisions. Moreover, certain trouble signals are cross connected between the two units so that if any additional time cycles are inserted in one unit for remedial action, the same number of cycles would be inserted in the other central control unit.

To provide this synchronous type of operation, two match circuits and associated registers are employed within each central control to make continuous checks of the operations between the two units. These checks perform repeated comparisons at certain points in one central control against corresponding points in the other unit for detecting mismatches. For this matching function, two binary data words (one from each central control) are compared in each 5.5-microsecond timing cycle. The binary words used for this matching operation are usually selected from the following processing points within central control depending on the particular program instruction:

data buffer register (DBR)
masked bus
unmasked bus
program address register (PAR)
buffer order word register (BOWR)
index adder
outputs of decoders and sequencing circuits

A simplified block diagram of the matching circuits and related components is presented in Fig. 4-10. Two internal match buses, labeled 0 and 1, provide access to selected processing points such as the program address register, data buffer register, index adder, and the masked and unmasked buses under control of a decoder in the match circuit. Sampled information from these points is transmitted over the internal match buses to the match registers under control of the gates, and simultaneously to the other central control unit via the external match buses 0 and 1, respectively, as indicated in Fig. 4-10.

Decoders in the match circuits in both central controls normally operate in step with each other. Therefore, the information transmitted from central control 0 to central control 1 is stored in the match registers in synchronization with the action of the gates. The two match circuits in each central control compare the contents of their associated match registers, and generate corresponding output signals in accordance with the presence or absence of a match condition. For example, the detection of a mismatch condition will generate an interrupt signal and further matching operations will be automatically stopped.

The selection of the points to be matched also facilitates the detection of hardware or equipment troubles developing within central control especially during routine tests made in accordance with maintenance programs. Thus, a mismatch condition is detected as soon as the effect of the trouble would be communicated to other elements in the electronic switching system. Signals from the decoders and sequence circuits within each central control also are sent to the respective match circuits as shown in Fig. 4-10. These signals set and reset specific flip-flops which, in turn, govern the selection of internal points within central control for matching during each 5.5-microsecond time cycle.

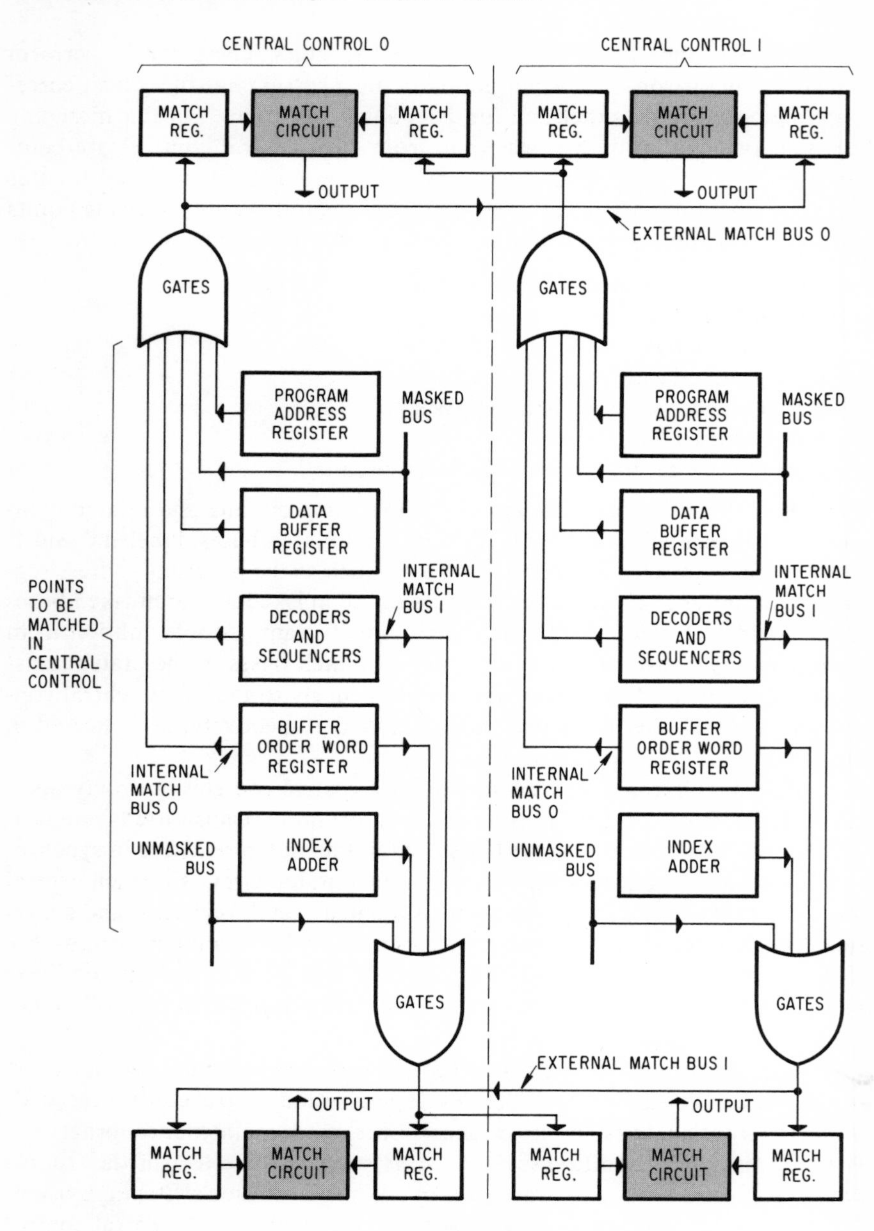


Fig. 4-10. Central Control Matching Circuits

## Bus Systems and Communications with Peripheral Units

In carrying out its functions, central control exchanges information (communicates) with its various associated components and peripheral units over common groups of twisted-pair wires called bus systems. Four major bus systems are employed for this purpose. They are designated program store bus, call store bus, central pulse distributor (CPD) bus, and the peripheral unit (PU) bus. Duplicated buses, labeled 0 and 1, link the two central control units with the program and call stores as shown in Fig. 4-11, in order to provide greater reliability.

A simplified block diagram of these major bus systems, which appears in Fig. 4-11, shows the flow of information between central control and the associated elements in the No. 1 ESS. One-half of the program stores and one-half of the call stores are normally installed to the left and right, respectively, of the two central control units in order to reduce the lengths of the respective buses.

The central pulse distributor (CPD) bus and the peripheral unit (PU) bus systems consist of two one-way groups of small gauge twisted-pair wires. The address group of these wire pairs handles the pulses and signals from central control to all units connected to these buses. The answer group of the

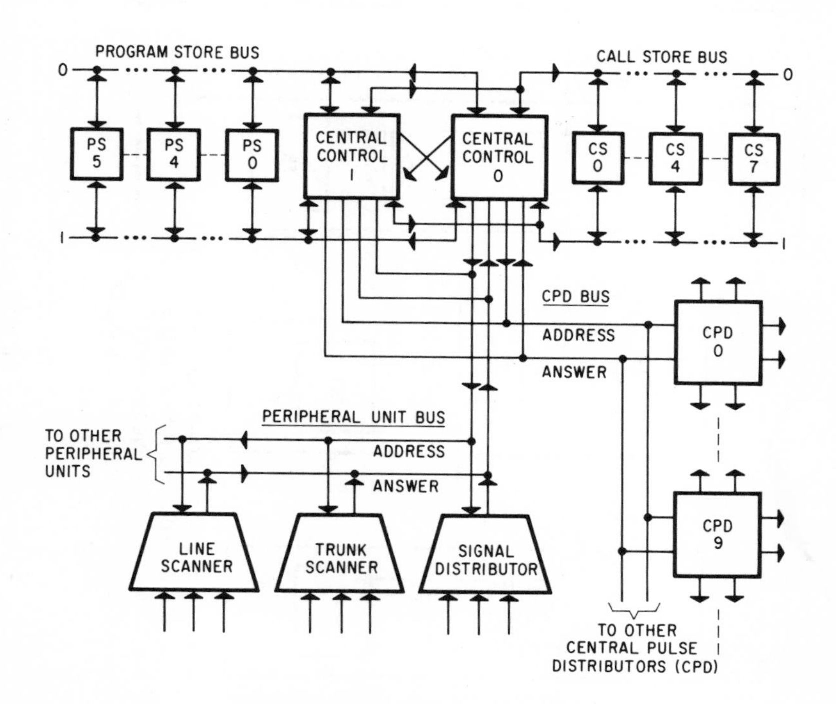


**Fig. 4-11.** Central Control Bus Systems

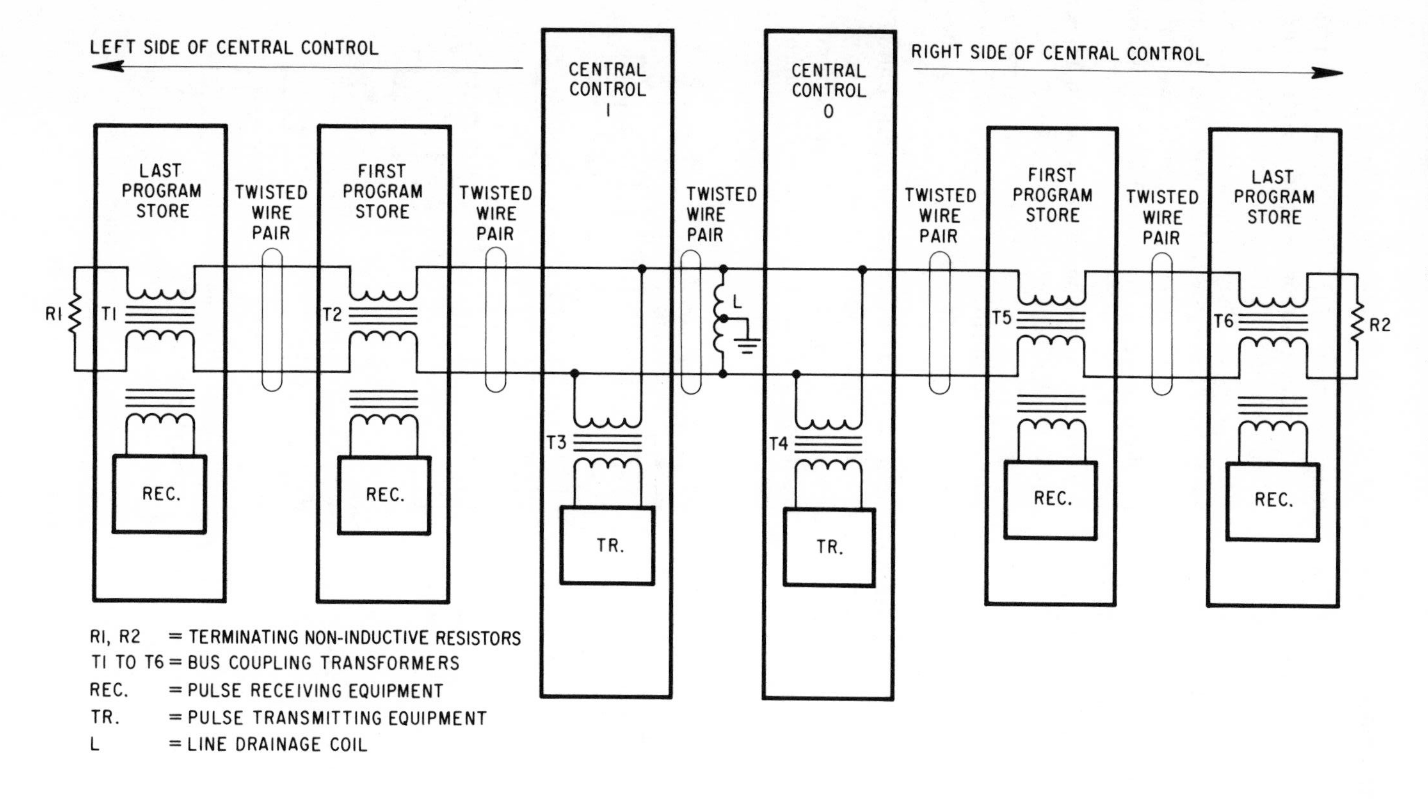


**Fig. 4-12.** Simplified Schematic of Typical Wire Pair Used in Program Store Bus

CPD and PU bus systems conducts the responses from selected peripheral units back to central control. One wire pair is provided in a bus for each binary digit or bit to be transmitted over the bus. A gating method permits a bus to be time-shared by the different units that it serves. The flow of information on any wire pair of a bus usually is in the same direction.

Pulses and other signals usually are applied to a bus every 5.5 microseconds and they may be only 0.5 microsecond in width. Therefore, in order to minimize propagation delays or transmission time over a bus, it is necessary that the maximum length of wire pairs between central control and the last program store or call store not exceed 125 feet. For the units on the peripheral unit bus and the central distributor bus, wire pair lengths up to approximately 450 feet may be employed without undue propagation delay effects. Note that a total of sixteen central pulse distributors connect to the CPD bus although only a quantity of ten is indicated in Fig. 4-11.

Each wire pair of a bus is connected to a bridged component of the bus through a transformer. Figure 4-12 is a simplified schematic diagram showing this transformer coupling arrangement for a wire pair in the program store bus. The pulse transmitting equipment in central control is connected to the primary of a transformer whose secondary is bridged across the wire pair. In each program store, the pulse receiver connects through a transformer whose primary windings are in series with the wire pair. Each end of a wire pair is terminated in a non-inductive resistor, indicated by R-1 and R-2 in Fig. 4-12, in order to maintain the proper line impedance and thereby reduce reflections of current pulses. The center-tapped inductance, L, is bridged across the wire pair to reduce the effects of induced noise by providing a path to ground.

Central control communicates with the sixteen central pulse distributors over the central pulse distributor bus as illustrated in Fig. 4-11. These central pulse distributors are utilized by central control to send orders to or obtain information from peripheral units, such as line scanners, junctor scanners, trunk scanners, network controllers, and signal distributors. For instance, assume that certain line scanning data is needed by central control from central pulse distributor 8. Referring to Fig. 4-13, central control first sends an enable address pulse over the address lead of the CPD bus to all central pulse distributors. Next, central control initiates an execute signal over an individual pair of wires to central pulse distributor 8 which responds by returning a signal over its echo lead. This return or echo signal permits central control to verify that the proper central pulse distributor (CPD 8) has been connected.

The selected central pulse distributor (CPD 8) will transmit an enable pulse over a separate direct path to the desired line scanner, 15, as illustrated in Fig. 4-13. This enable pulse causes line scanner 15 to respond to the address signal subsequently transmitted by central control to all peripheral units

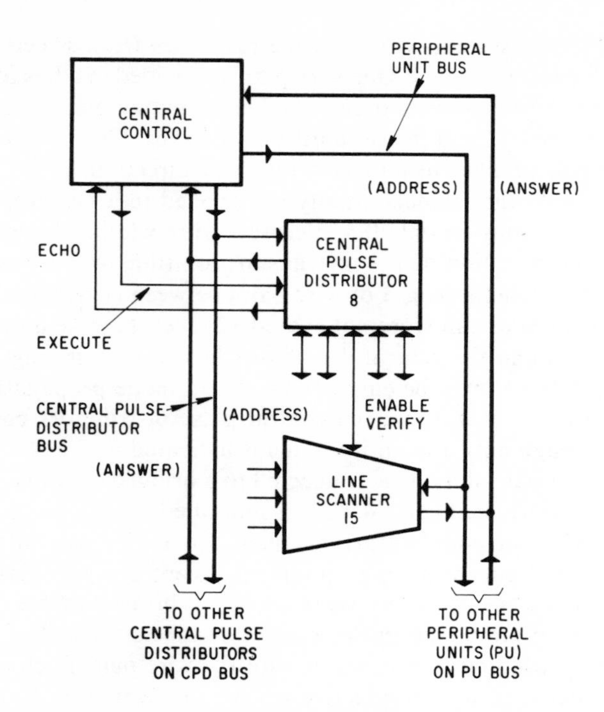


**Fig. 4-13.** Applications of CPD and PU Bus Systems

over the address lead of the peripheral unit bus. Line scanner 15 acknowledges by sending a verify signal back over its same connecting path to central pulse distributor 8. This verify signal is translated by CPD 8 into coded binary digits which are transmitted to central control over the answer lead of the central pulse distributor bus. In this manner, central control can match the returned binary digits from CPD 8 with the original address signal that was previously transmitted in order to determine that only the desired line scanner 15 had been activated.

The information that is sent to a particular central pulse distributor, containing the address of a required peripheral unit, such as line scanner 15, is set up in the F register by appropriate instructions received from program store. The data in the F register is then sent to the central pulse distributor address translator which transmits the proper pulses to the central pulse distributors over the CPD bus as previously explained. The verify answer signal returned from the selected central pulse distributor, for example, CPD 8, is gated into the Y register. This signal in the Y register is then compared in a match circuit with the data in the central pulse distributor address translator. If a mismatch results, an interrupt signal will be generated and a program transfer will be initiated by central control as previously explained.

Information for controlling a particular peripheral unit, such as line scanner 15, is placed in the K addend register and in the K register. The data in the K addend register is decoded into the required binary form by one of the peripheral address translators before being transmitted to line scanner 15. The readout from this line scanner, which is returned over the answer lead of the peripheral unit bus, is gated into the L register for further processing by central control.

## Signal Processor Interconnection

In the larger electronic switching offices, a signal processor (SP) often is provided to assist central control. A signal processor may be defined as a stored-program processing unit that can efficiently handle input-output (I/O) functions. For instance, the signal processor can detect originating calls, receive dial information from calling customers and process the calls, control the peripheral units of central control, etc. By relieving central control of the aforementioned tasks, the signal processor can increase substantially the traffic capacity of an electronic central office. Two pairs of duplicate signal processors may be provided depending on the size of the central office and its traffic requirements.

The signal processor, in a similar manner as central control, operates under the direction of program instructions. These program instructions, however, are kept in call stores that are associated only with the signal processor. Figure 4-14 shows the relationship of the signal processor and its associated call stores with central control. Up to eight signal processor call stores may be connected with each pair of duplicate signal processors.

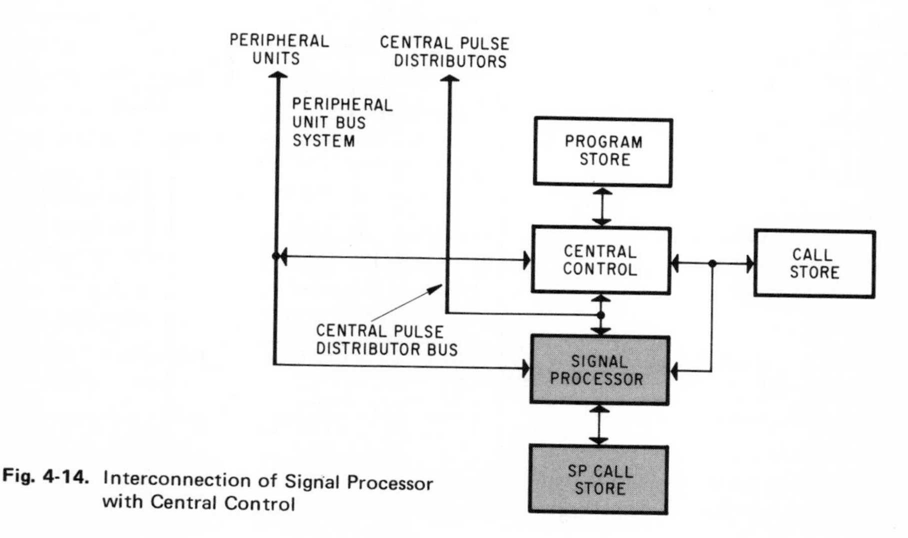


**Fig. 4-14.** Interconnection of Signal Processor with Central Control

Central control and the signal processor operate independently of and simultaneously with each other. Central control can always address an SP call store as if it were another call store on its call store bus system. The tasks performed by the signal processor include supervision or periodic scanning of lines to detect originating calls, junctors to detect disconnects, and trunks to check for connections and disconnects. Other actions include the reception of dial pulses or Touch Tone® signals from customers, the reception and transmission of signals with other central offices, and various types of timing operations to detect permanent signals and other malfunctions.

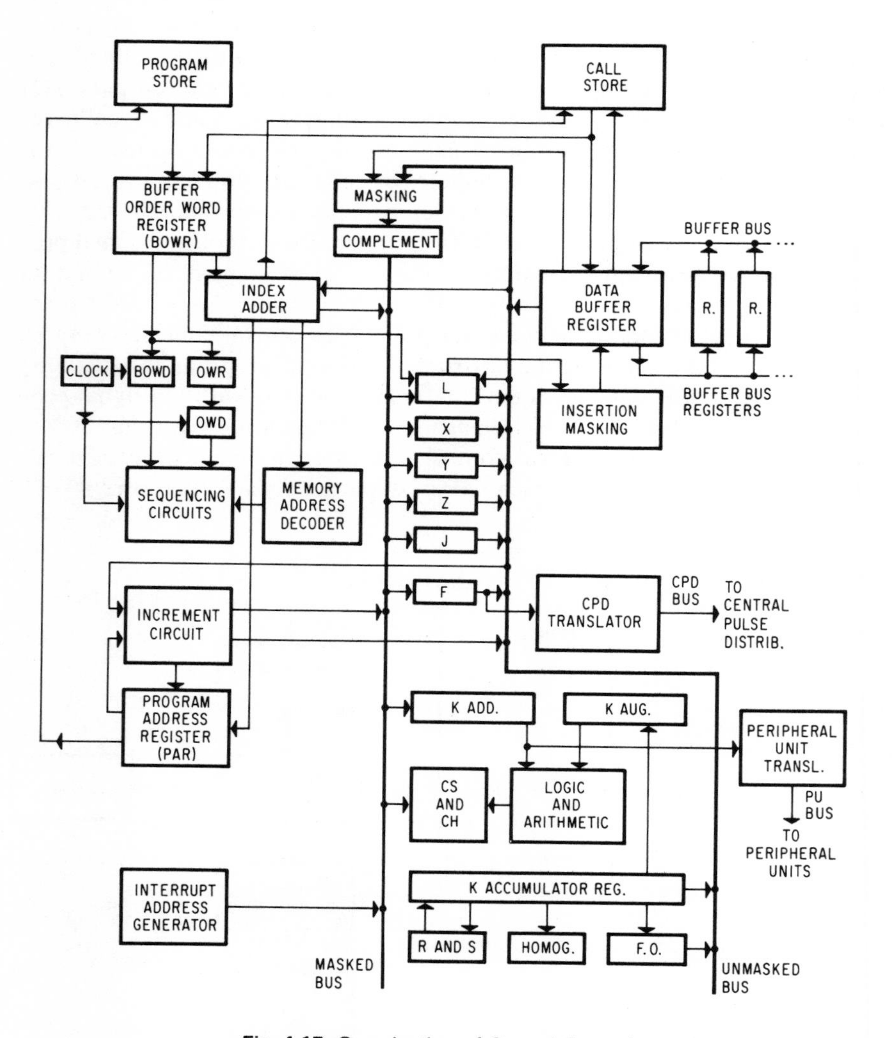


**Fig. 4-15.** Organization of Central Control

## Summary of Central Control Organization

Many of the principal characteristics and functions of central control, including its major components, have been described in this chapter. The basic organization of central control also has been explained by a series of step-by-step block diagrams depicting its major components and subsystems. In addition, some aspects of the logic design of central control are described including program instruction operations and transfers, data processing, timing requirements and overlap operations, use of sequencing and matching circuits, and the major bus systems for communicating with associated elements and peripheral units.

Figure 4-15 is a simplified and concise drawing showing, by block diagrams, the major components and associated elements of central control that are explained in this chapter. Since it is not feasible to delineate in this drawing all of the components and their functions, the reader should refer to the text and other related drawings for more detailed information.

## Questions

1. What three principal parts comprise the central control unit? What components interface between central control and program store? Between central control and call store?
2. What are the main functions of: The buffer order word register (BOWR)? The program address register (PAR)? The data buffer register (DBR)? The index adder?
3. Name four or more index registers that are mainly concerned with the processing operations of central control. What buses connect to these index registers?
4. What circuits within central control provide masking functions? Explain how masking is performed.
5. Explain the meaning of the following instruction received from program store by central control: "MZ3X."
6. Name three or more functions performed by sequencing circuits.
7. What general functions are performed by the K and J registers? By the CS and CH flip-flops?
8. What are the functions of interrupt signals? How are they originated?
9. What operations are carried out by the match circuits for the two central control units? From what three or more processing points within central control are binary words normally selected for this matching operation?
10. Name the major bus systems used by central control for communicating with its major elements. Describe a typical bus system.

# 5
# MEMORY SYSTEMS AND DEVICES

## Types of Memory Systems

The general functions of central control in processing calls, as explained in Chapter 4, are directed by program instructions and translation information contained within a semipermanent memory system called program store. In accomplishing the various call processing tasks, central control also utilizes a temporary memory unit, termed call store, to briefly store and then recall or retrieve data pertaining to the called number and its routing. The general principles and functions of program store and call store memory systems are explained in this chapter.

The instructions and other data stored in these two memory systems are in the form of binary data words. The semipermanent memory employed in program store is so named because data can be read out or recalled from it only by central control. It is necessary that external means, as later described, be used to record or write the required program instructions and other information to be kept within program store. On the other hand, call store has erasable memory storage facilities for temporarily storing data relating to a call in progress. Thus, central control may write binary data into call store for temporary storage and subsequent recall or read out during the progress of a call.

In addition to the Bell System's ESS No. 1 and No. 2, other electronic switching systems such as the General System's No. 1 EAX and equivalent types utilize semipermanent and temporary memory systems. These different memory arrangements are designed to store a binary digit or bit in a specific location in the memory device and to immediately recall it on request. Likewise, it is possible to store several millions of bits in a single memory module such as used in program store. Some of these bits may be stored for years and recalled as often as needed. In call store, the bits are temporarily stored and read out during the course of a call, and then erased when new data for a succeeding call is received.

## Basic Memory Cell

The basic element of the memory cell employed in both program store and call store is a ferrite magnetic core. This magnetic core consists of a bistable ferrite material of uniform composition which possesses a square-loop magnetic characteristic as shown in Fig. 5-1. It can be magnetized in either a clockwise or counterclockwise direction. But, more important, the magnetization of this type ferrite core can be switched very rapidly from one direction to the other.

For instance, assume that the ferrite core is magnetized in a positive or clockwise direction toward point M on the square-loop curve as indicated in Fig. 5-1. It will retain this positive magnetization after the removal of the magnetizing force. To switch or reverse the direction of magnetization, that is, from M to N or in a counterclockwise direction, it would be necessary to apply a negative magnetizing force that can overcome the positive threshold value indicated by R in Fig. 5-1. Similarly, if the core had been magnetized in a negative or counterclockwise direction, it would be necessary to apply a positive magnetic drive to exceed the negative threshold value of S in order to switch the ferrite core's magnetization from the N to the M value shown in Fig. 5-1.

Many thousands of these ferrite cores are assembled in a coordinate frame of wires to form a matrix or memory module as used in program store. In order to understand the functions of these magnetic ferrite cores in a memory system, let us consider a simple arrangement of a number of these ferrite cores as illustrated in Fig. 5-2. Note that two wires intersect at right-angles in the X and Y directions (orthogonally) in the center of each core. The vertical or X column wire is called the X selection wire and the horizontal or Y row

**Fig. 5-1.** Square-loop Characteristics of a Ferrite Core

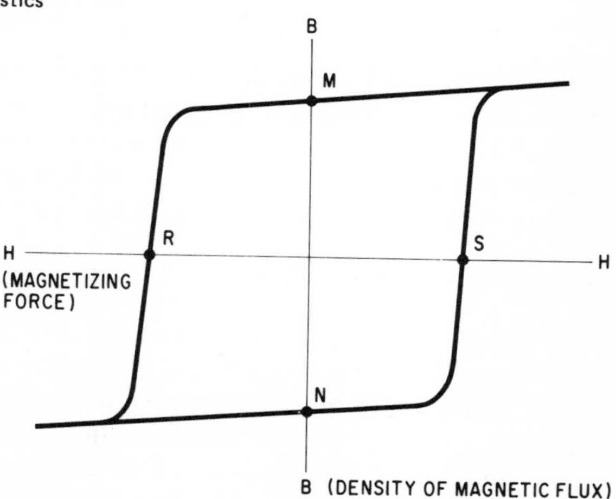

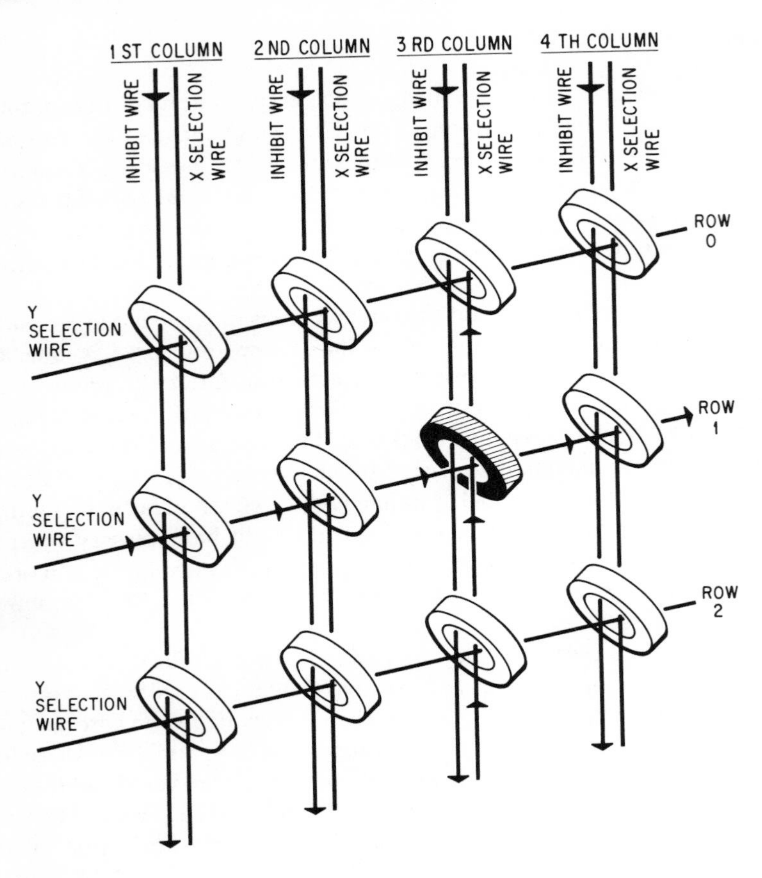


**Fig. 5-2.** Elements of Basic Memory Cell Utilizing Ferrite Cores

wire is called the Y selection wire. Another wire, termed the inhibit wire, passes vertically through the core parallel to the X selection wire. This wire has a continuous direct current flowing through it as a bias current. This bias current serves to restore the core to its initial magnetic state after the removal of the current pulses from the X and Y selection wires.

## *Selection of Memory Cell*

It is essential that direct-current pulses be applied in the required directions through the intersecting X and Y selection wires in order to select a particular memory cell for subsequent readout of its stored binary digit. Consequently, only one-half of the current value needed to change the magnetization of the ferrite core is applied to the X column wire and the other half is sent through the Y row wire. The current in each wire, as a result, will be less than the amount required to change the core's magnetization, for instance, from clockwise to counterclockwise or vice versa. This arrangement prevents the magnetization of a core from being changed if only one of the selection wires passing through it carries current. It also insures that only the particular ferrite core at the intersection of the current-carrying X column and Y row wires will be magnetized or have its magnetization changed after the current pulses are removed.

Referring to Fig. 5-2, observe that the Y selection wire in the second row has current flowing in the left to right direction. At the same instant, the X selection wire in the third column (going from left to right) has a current pulse in the upward direction. Thus, only the ferrite core located in the third column of the second row will be selected because both X and Y current pulses flow through its core windings. The other ferrite cores in the third column have only X current flowing in the wires through them and only Y current is flowing through the other ferrite cores in the second row. Consequently, there will be insufficient current to magnetize these ferrite cores. The selected core, shown shaded in Fig. 5-2, is magnetized in a clockwise direction because of the direction of the currents flowing through its intersecting X and Y selection wire.

### *Storage and Readout Operations in Memory Cell*

Assuming that the ferrite core of the selected memory cell was magnetized in a clockwise direction, a binary digit 0 would be stored or written into this particular cell. If the core had been magnetized in a counterclockwise direction, a binary digit 1 would be stored instead. This same principle is applicable to the other memory cells in the modules contained in program store and in call store. The ferrite core, however, is only one of the elements of the modular memory in program store. Other memory devices are utilized in both program store and in call store as later described.

To recall or read out a binary digit that had been stored in the memory cell depicted in Fig. 5-2, the same current pulses are simultaneously applied to the X and Y select wires that intersect in the selected ferrite core. If this core had been previously magnetized in the same (clockwise) direction that would be induced by these current pulses, the core's magnetization will not change. Therefore, no voltage would be induced into the readout loop or sensing wire associated with the memory module. However, if the applied current pulses are in the opposite direction, the core's magnetization will be switched. This reversal of the core's magnetic flux will cause a voltage pulse to be induced into the readout loop. Therefore, to recall or read out binary digits from memory cells may be likened to checking for voltage and no-voltage conditions.

To store or read out from a simple memory cell only one bit at a time is not practical for the high-speed requirements of electronic telephone switching systems. Consequently, bits are arranged together in parallel combinations called words so that many bits may be stored in or recalled from the memory module simultaneously. It is possible to form binary words of almost any number of bits by connecting the ferrite cores in certain matrix or wiring patterns. For instance, in the program store memory of the No. 1 ESS central office, each binary word contains 44 bits. In call store, the binary words have 24 bits as later explained.

## Organization of Program Store

The memory elements used in the No. 1 ESS program store, in addition to the ferrite cores previously described, include a series of twistor wire pairs which are used for readout purposes, copper-strip solenoid loops, and a stack of aluminum memory cards (each approximately 6½ inches by 11½ inches). These elements are assembled to form the basic memory module of program store. Each aluminum memory card contains 2,880 tiny rectangular or bar magnets made of vicalloy, a special magnetic alloy similar to that used in tape recorders. These minute permanent bar magnets on the aluminum cards are arranged in a grid of forty-five columns and sixty-four rows as illustrated in Fig. 5-3. Every row has forty-five miniature bar magnets of which forty-four are used to represent one binary data word (the last or forty-fifth magnet is not utilized). Each aluminum memory card, therefore, has a capacity of sixty-four binary words (each of forty-four bits) or a quantity of 2,816 binary digits or bits.

A total of 128 of these aluminum memory cards are stacked together with their associated cables of twistor wire pairs, copper-strip solenoid loops, and ferrite access cores to form one memory module which can store up to 8,192 binary words. A package of sixteen memory modules, containing an aggregate of 131,072 binary words (approximately 5,800,000 bits) makes up one program store unit. Depending on the number of lines, traffic requirements, and related factors, from two to six program stores usually are installed in a No. 1 ESS central office.

The 131,072 binary words in the memory field of a program store are divided into two equal groups or blocks designated H and G. (The "G" and "H" designations were derived from mule driver's terminology, "Gee" and "Haw," for right and left, respectively.) The H half is on the left and the G half is on the right when looking at the front of a program store frame. To ensure the availability of program instructions in the case of power failure, bus or other malfunctions, the H and G halves of a program store are installed on different program store frames. In this way, full duplication of the program memory is provided in an ESS central office.

As shown in Fig. 5-4, the program instructions in the H half of program store 1 are duplicated in the G half of program store 0, and the H half of program store 0 is duplicated in the G half of program store 2. Thus, duplicate

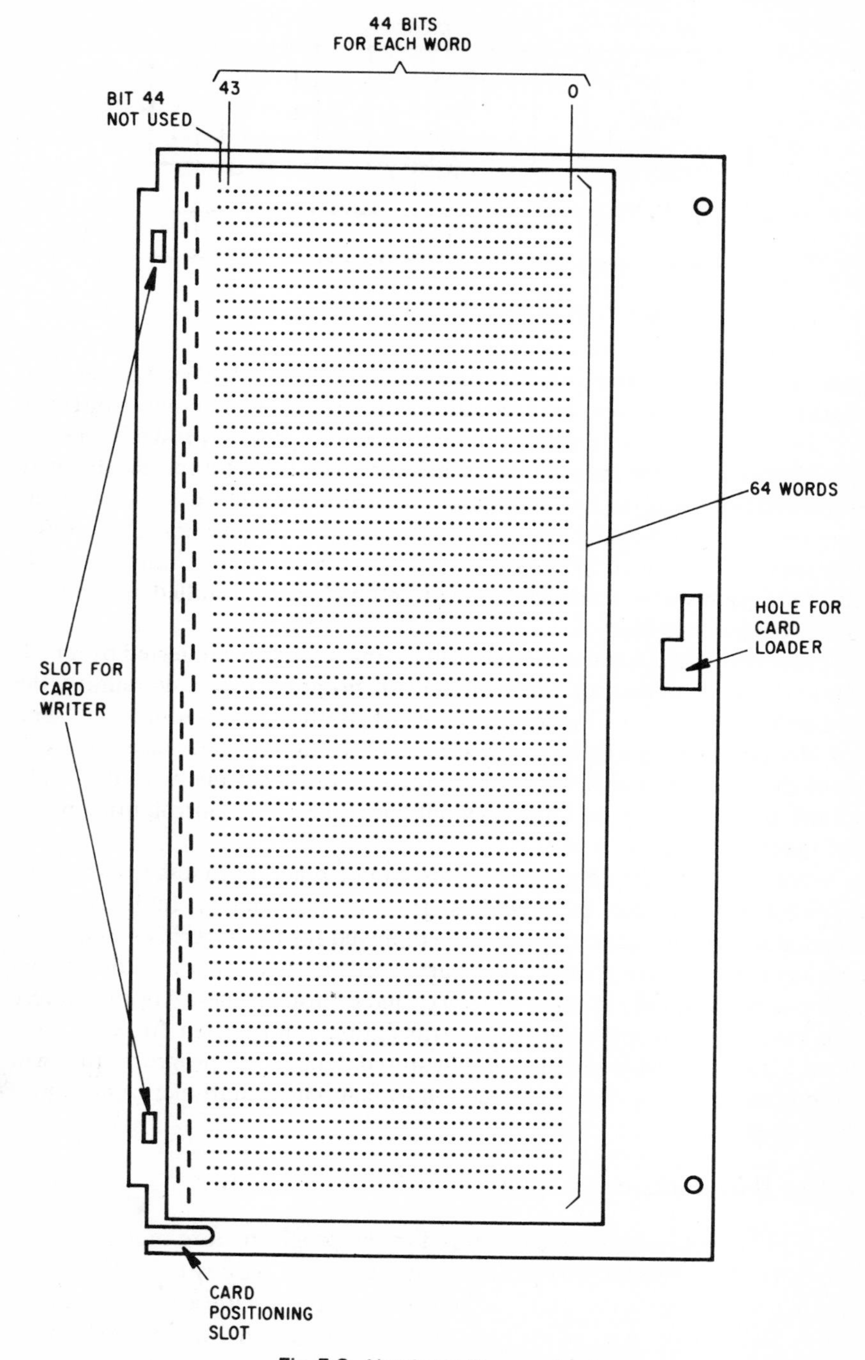


**Fig. 5-3.** Aluminum Memory Card

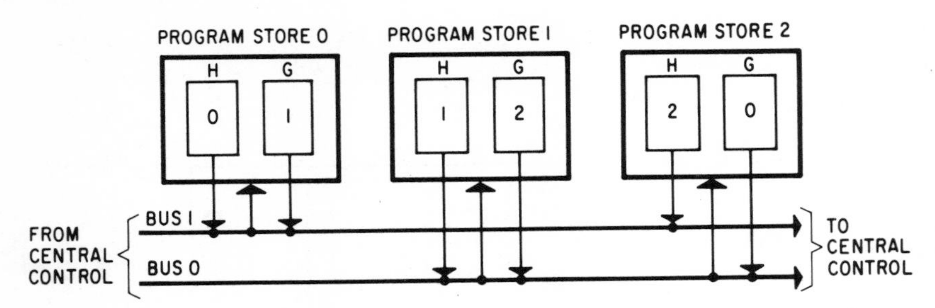


**Fig. 5-4.** Duplication Method for Program Stores

program instructions and information always will be available in the adjacent program store. Moreover, an ESS office can function with an odd number of program stores and also provide full duplication of program instructions.

The H half of a program store memory generally communicates with central control over one bus and the G half uses the other bus. For instance, referring to Fig. 5-4 again, the H half of program store 1 connects to bus 0 while its G half, located in program store 0, utilizes bus 1. Similarly, the H half of program store 2 connects to bus 1 and its G half, installed on program store 1, connects to bus 0.

Consequently, whenever a particular binary word is requested by an address from central control, each of the two program stores containing the word will be requested to read out the word. Both program stores will send back the binary data word to central control, using separate buses. Thus, if one of the program stores should malfunction, the other one with the duplicate information would be able to send it back to central control, utilizing either one or both buses.

Any binary word in program store can be read out in less than 5.5 microseconds upon receipt of its address from central control. In general, approximately 100,000 binary words are employed for the generic or basic program instructions. The number of binary words needed for translation information depends on the number of trunks and customer lines. Approximately three binary words are required as translation information for each line. Means also are provided to ensure that stored data within program store will be protected from accidental destruction by equipment malfunctions or operator errors.

## *Twistor Wire Pairs and Copper-Strip Solenoid Loops*

The readout of binary data stored in the program store modules is accomplished by a *twistor wire pair or loop.* A twistor wire is a length of very small gauge copper wire which is helically wrapped with a thin permalloy tape. A plain copper wire of the same gauge is run parallel to the twistor wire

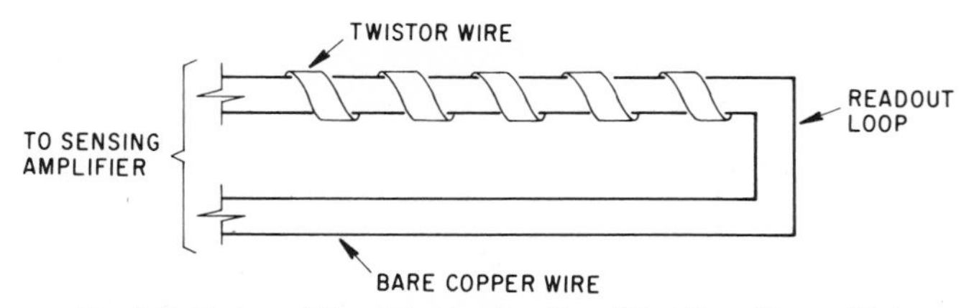


**Fig. 5-5.** Twistor Wire Pair Used to Read Out One Binary Digit

and connected to it at one end to form a readout pair or loop. The other end of the pair connects to a sensing amplifier and other circuits in program store required for readout purposes. Figure 5-5 illustrates the basic twistor wire loop used in the memory module. A total of forty-five twistor wire pairs are assembled into a flat plastic cable. Two such cables are utilized for each memory module, one on each side of the solenoid loops as described below.

The main part of the memory module consists of a vertical stack of sixty-four *copper-strip solenoid loops,* each of which passes through the center of a small ferrite core attached to an insulating board. The two aluminum memory cards face this stack of sixty-four copper-strip solenoid loops, one card on each side of the stack. Two forty-five-pair twistor wire cables are installed perpendicular to the respective sides of the solenoid loops so that each cable is between the stack of sixty-four solenoid loops and an aluminum card. Figure 5-6 is a simplified sketch of this assemblage.

A twistor wire pair, therefore, will traverse the same binary digit or bit position in every one of the sixty-four rows of binary words represented on an aluminum card; that is, one of the miniature bar magnets on each aluminum card will appear at every intersection of a twistor wire pair and a copper-strip solenoid loop. Since one aluminum memory card appears on each side of a stack of solenoid loops, the sixty-four insulating boards, with their stacks of solenoid loops that comprise a memory module, are used to read out binary words on the 128 aluminum cards.

## *Selection of Ferrite Access Core*

A complete memory module in a program store, as previously described, consists of sixty-four separate sections of solenoid stacks. Every section contains a stack of sixty-four solenoids and ferrite access cores. Consequently, there will be a total of 4,096 ferrite cores in a memory module arranged in a sixty-four by sixty-four matrix or array. The sixteen memory modules in one program store make up 65,536 ferrite access cores.

The two common vertical wires, the bias and X selection leads, and the Y selection lead wire run through the core at right angles to each other. The X and Y wires actually form two turns around the ferrite core. The single bias wire has a direct current continuously flowing through it.

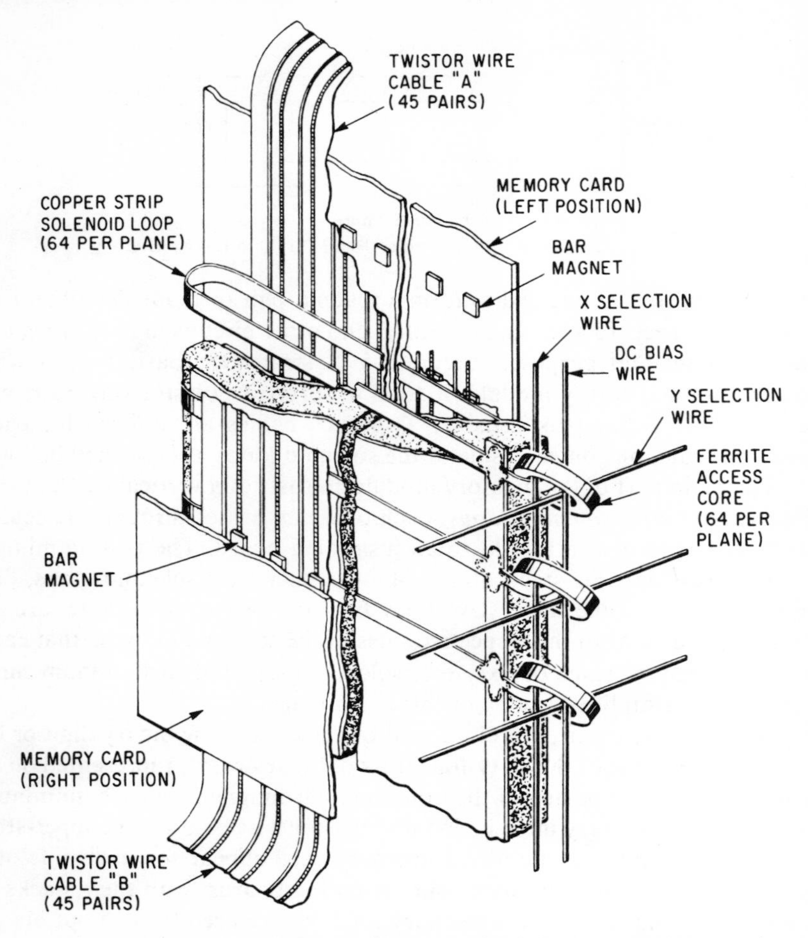


**Fig. 5-6.** Memory Module Section (Courtesy of Bell Telephone Laboratories)

To initiate a readout of a specific binary word from the sixty-four words stored on each of the 128 aluminum memory cards positioned in a memory module, it first is necessary to select the proper ferrite access core. This selection is governed by the address information received from central control in the form of binary digits, via the program store's control circuits.

For the selection process, the H and G halves of each program store are assigned a four-digit binary name. This name contains four binary digits, two binary ones and two binary zeros, for example, 1001. In addition, a 16-bit address code is used to identify the desired binary word from the 65,536 binary words stored in each H and G half of a program store. These binary digits indicate the location of the desired word by its particular memory module and H or G designation. This address data also is sent from central control.

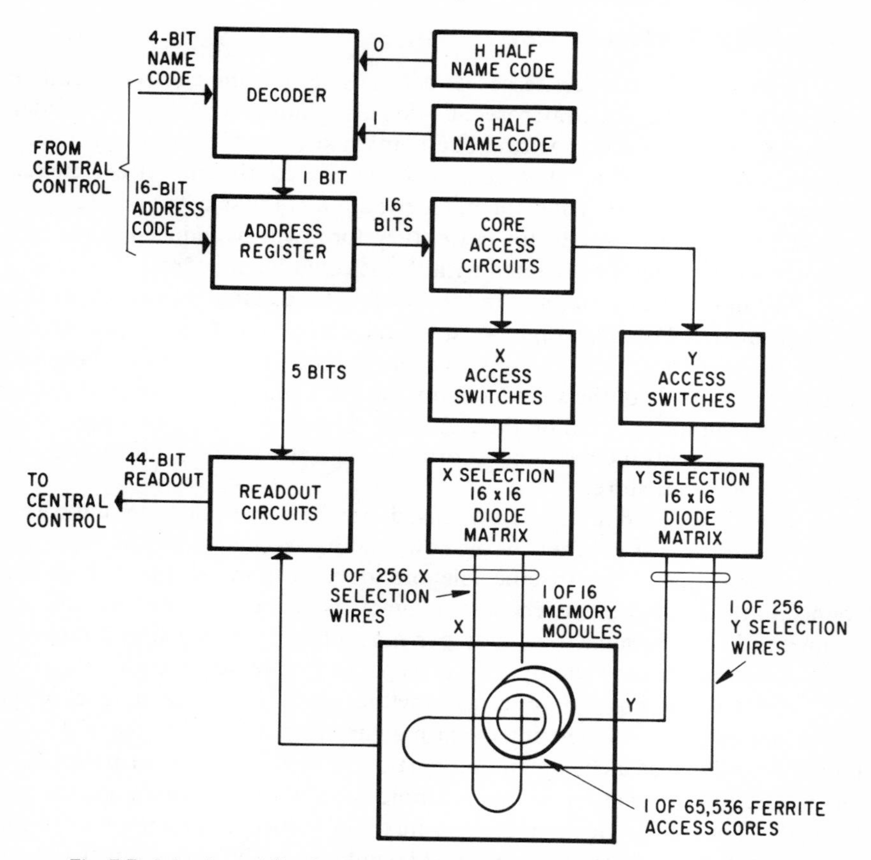


**Fig. 5-7.** Selection of Ferrite Access Core in Memory Module of Program Store

The main elements of a program store's control and core access selection circuits are shown in the block diagram of Fig. 5-7. The four binary digits of the name designation are compared by the decoder unit with the related binary digits assigned to the H and G halves of program store. If there is a match, the decoder will generate a binary 0 digit for the H half or a binary 1 for the G half. The particular binary digit or bit is immediately sent to the address register. Therefore, seventeen bits will be contained within the address register.

The address data recorded in the address register will cause the core access circuits to connect to the proper X and Y access switches. These switches, in turn, will select the required X and Y diode matrixes in order to determine one of the 256 X selection wires and one of the 256 Y selection wires which will intersect in the desired ferrite access core. This process is illustrated in Fig. 5-7.

## Binary Digit Readout Process

When the desired ferrite access core is selected, the control circuits in program store initiate a 2-microsecond dc pulse simultaneously over its intersecting X and Y selection wires. The combined X and Y pulses produce a magnetic field exceeding that generated by the continuous direct current flowing in the bias wire. Thus, the magnetic state of the selected ferrite access core will be changed. A current pulse, therefore, will be induced in the copper-strip solenoid loop associated with the selected ferrite core.

The induced current pulse in the solenoid loop actually causes the interrogation of two 44-binary digit words at the same time. One binary word is stored on the adjacent row of forty-four bar magnets located in the aluminum card on the left side of the solenoid loop; the other binary word is on the same row on the aluminum card on the right side. However, only one binary word will be sent to central control in accordance with the address information received by program store.

The binary readout process involved may be better understood if we examine one of the bar magnets in the row, and its associated twistor wire pair at the intersection of the specific solenoid loop, as shown in Fig. 5-8. In this connection, recall that each bar magnet is either magnetized or demagnetized. A magnetized vicalloy bar magnet represents a stored binary 0 digit. A demagnetized vicalloy bar corresponds to a binary 1 digit.

Referring to Fig. 5-8, let us assume that the first vicalloy magnet in the row is not magnetized. The interrogating current pulse in the solenoid loop, therefore, will cause a change in that part of the twistor wire's permalloy tape located at its intersection on the solenoid loop with the unmagnetized bar magnet. This change in the magnetic flux will induce a voltage pulse in the twistor wire. No voltage, however, will be induced in the plain copper wire of this twistor wire pair. This plain wire provides a return path for the readout

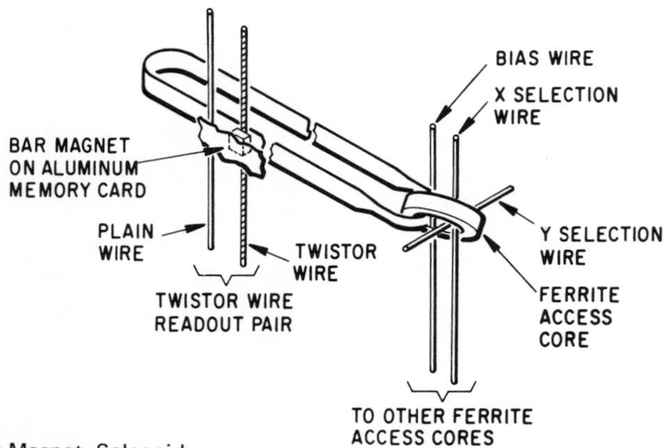


**Fig. 5-8.** Intersection of Bar Magnet, Solenoid Loop, and Twistor Wire Readout Pair

current through the external connected circuits. Consequently, the induced voltage in the twistor wire pair will result in a binary 1 readout for the unmagnetized bar magnet in this row on the aluminum memory card. Moreover, any induced noise voltage will be in the same direction in both the twistor and plain wires of the readout pair and, therefore, will cancel out.

When the current pulse is removed from the X and Y leads, the specific ferrite access core at their intersection will be restored to its initial magnetic state by the continuous direct current in the bias wire. This action will produce a current pulse in the respective solenoid loop which will be opposite in direction to the initial interrogate pulse. Thus, the portion of the permalloy tape at the intersection of the twistor wire and the solenoid loop will change back to its original magnetic polarity.

Let us now assume that the next vicalloy bar magnet in the same row on the aluminum card is permanently magnetized. In this case, the magnetic field caused by the interrogating current pulse in the solenoid loop will add to the magnetic field of this magnetized bar magnet. Therefore, there will be no change in the magnetization of the twistor wire opposite this bar magnet. As a result, little if any voltage will be induced in the respective twistor wire pair so that the readout will be a binary 0.

At the completion of the current pulses on the X and Y leads, the ferrite access core will also revert to its original magnetic state. The resultant induced voltage pulse in the solenoid loop now will generate a magnetic field which is opposite to, but weaker than, the magnetic field due to the permanent vicalloy bar magnet. As a result, the permalloy tape of the twistor wire pair will retain its initial magnetic polarity.

## Readout of Binary Word from Program Store

The foregoing readout process is applied simultaneously to the forty-four operative vicalloy bar magnets on the same row of both aluminum cards facing the chosen solenoid loop. Therefore, either a binary 0 or a binary 1 signal will ensue in each of the forty-four associated twistor wire pairs. The simultaneous readout of forty-four binary digits comprises one binary word.

The method of handling the binary word readouts can be seen by referring back to Fig. 5-6. The forty-four twistor wire pairs in cable A will convey the 44-bit binary word readout from the left aluminum memory card. In a similar manner and at the same time, the different binary word readout from the aluminum card on the right side of the solenoid stack will be carried by the forty-four twistor wire pairs of cable B. Selection circuits in program store, as indicated in Fig. 5-9, are utilized to select and amplify the desired binary word in accordance with the address information received from central control.

To reduce circuit complexities and the number of individual amplifiers required, the readout leads of the sixteen memory modules in a program store

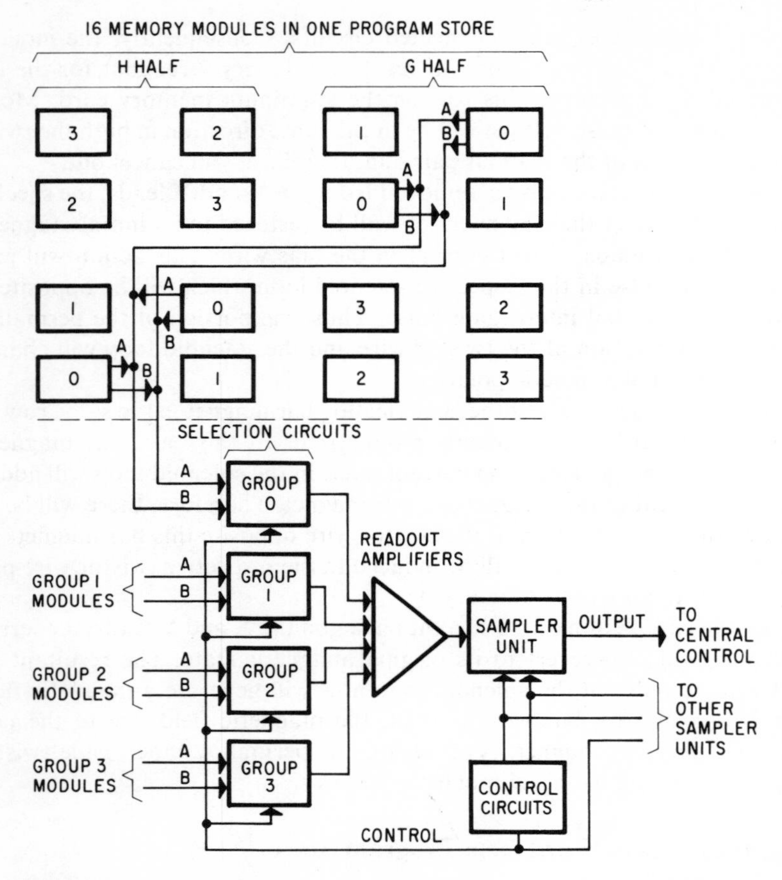


**Fig. 5-9.** Readout Selection and Connection Circuits for Memory Modules in Program Store

are paralleled in groups of four. Each of the four modules in a group is picked from a different location in program store to minimize electrical noise interference. Figure 5-9 is a block diagram indicating a typical division of the sixteen memory modules into four groups, designated 0 to 3, inclusive.

The readout pairs of the four similar numbered modules in a group are connected in parallel as shown for the modules designated 0 in Fig. 5-9. Groups numbered 1, 2, and 3 are connected in a similar fashion. Thus, four groups of eighty-eight readout pairs each are connected to the selection circuits where the desired group of forty-four readout pairs is chosen.

A group of forty-four readout amplifiers provides the necessary gain to the outputs of the selection circuits (as shown in Fig. 5-9) in order to properly operate the sampler unit. The sampler unit makes the decision whether the output is a binary 1 or a binary 0. It also quantizes or prepares the output pulse to be a length of one-half microsecond for subsequent transmission over the connecting bus to central control. The output of the sampler, therefore, will contain a series of forty-four binary one and zero signals which form the desired binary word of the program instruction for use by central control.

## Changing Information in Memory Cards

There are, as previously explained, 2,048 aluminum or twistor memory cards in a program store, divided into 16 modules of 128 cards each. These memory cards contain the program instructions and translation information for directing the operation of central control and the ESS office. There is seldom a need to change the program instructions after an electronic central office is placed in operation. Translation information, such as trunk routings and status of customer lines, however, requires frequent modifications. Some of this information, relating to recent changes in customer lines and trunks, is temporarily stored in call store prior to updating the data in program store. Central control will refer to the translation information in program store if the desired information is not found in call store. It is necessary, usually at monthly intervals, to prepare a new set of twistor memory cards which will incorporate changes that have been made in the translation information.

The equipment used (see Fig. 5-10) to effect translation changes in program store consists of a memory card writer and two card loaders together with a set of 128 spare cards. The 128 twistor memory cards in a module are handled as a unit for effecting the translation changes. The cards are withdrawn from, or inserted into, a memory module by one of the motor-driven card loaders which also is used to transport them to the card writer unit. Recall that the sixty-four right-hand and the sixty-four left-hand twistor cards in a module have their magnets facing each other. Consequently, the magnets in one-half of the cards will be face up and the magnets in the other half will be face down. In order that all cards will be positioned so that their magnets will face up during the information writing process, two mounting positions are provided on the card writer unit, designated Pass A and Pass B.

To update translation information in a program store, a memory module is first selected in accordance with information received from central control over the teletypewriter at the master control center. The card loader, containing the spare set of 128 twistor memory cards, is mounted on the Pass A position of the memory card writer and the appropriate keys are operated to initiate the writing operations. The memory card writer unit obtains the new translation information from central control, one binary word at a time as needed for the writing process, and stores it in a register circuit. The card writer mechanism next withdraws one memory card at a time from the card loader for the writing process.

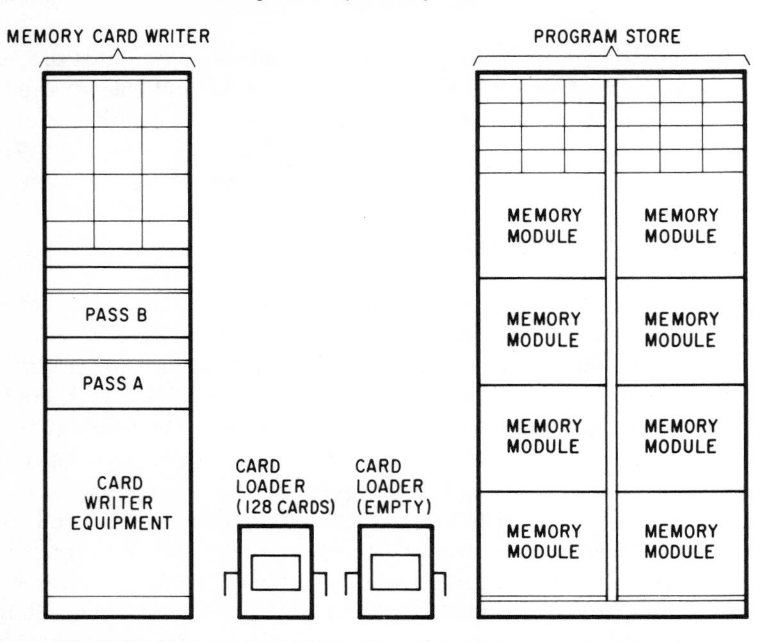


**Fig. 5-10.** Memory Card Writing Equipment Used with Program Store

The new translation information is written on the memory card, one 44-bit word at a time, by passing a special type of writing head across the surface of the card. A 20 kHz sine wave current is sent through a winding on the writing head. The miniature bar magnets in the memory card are magnetized or demagnetized by altering the waveform of the 20 kHz current in accordance with the binary data that was stored in the register circuit of the memory card writer. Upon completion of the writing process, which takes a matter of seconds, the memory card is returned to the card loader.

The writing process continues until all sixty-four right-hand cards have been completed. Then the card loader is manually inverted to the Pass B position and the remaining sixty-four left-hand cards are processed in the same manner. Upon completion of the card writing work on the 128 cards, a program store in which a memory module is to be updated is removed from service. The memory cards are removed from this module by the second card loader and the newly written cards in the first card loader are immediately inserted in the module. The translation information on the new cards is then verified with central control and, if found to be correct, the program store is returned to normal service. This sequence of steps is repeated to rewrite the cards in the memory module containing the duplicate of the particular translation information. The entire process is repeated as required for the other program stores.

## Call Store Memory Systems

The call stores furnish the temporary read-write or erasable memory systems in the No. 1 ESS central office. Unlike program store, information may be temporarily recorded in call store and then read out, altered, or erased as directed by central control. The information stored in the temporary memory of call store relates primarily to the handling of calls in progress within the electronic central office.

The type of information includes, for example, the network terminations used for each call in progress, the dialed digits received for the called number, the specific digits to be outpulsed or sent to the distant called office, the busy or idle status of customer lines, trunks, junctors and network links, customer billing information to be recorded on the automatic message accounting (AMA) tape, and recent changes in data concerning customer lines and trunks prior to updating the associated translation information in program store.

Duplicate input and output bus systems are provided to ensure high-speed reliable communications between call store and central control. These bus systems also control high-speed switching of duplicated memory blocks within call stores. Each call store communicates over either or both duplicated buses as directed by central control. All call stores share the same set of signaling and addressing leads. Furthermore, special maintenance control modes are provided for the interrogation of various test points within call stores, and to permit the alteration of operational features of internal circuits as may be required by central control.

## Organization of Call Store

Information stored in a call store is contained in binary data words of twenty-four bits each. One of these bits, however, is utilized for parity checking purposes. Each binary word has a specific location in call store which is identified by an address sent from central control. The particular mode to be performed, for instance, writing or readout, is specified by central control. For a writing mode, central control also sends the binary word to be written in call store. In case of a readout request, the required binary word is located and read out for transmission to central control. The interval between consecutive operations in call store is 5.5 microseconds.

There are 8,192 binary word locations in one call store. This provides a storage capacity of 196,608 bits on the basis of twenty-four bits per binary word. The number of call stores installed in an electronic central office depends on the number of customer lines and the busy-hour calling rate. For example, a 10,000-line central office with a busy-hour calling rate of 13,000 calls would usually require four call stores including the two duplicate ones. Large central offices may require as many as thirty-six call stores. If signal

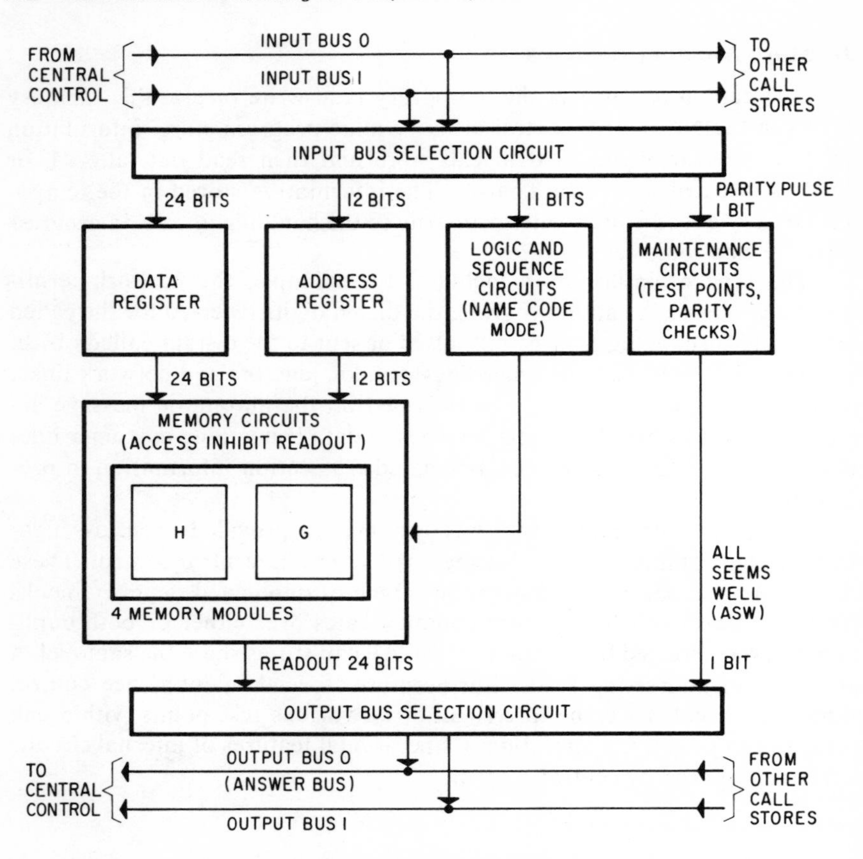


**Fig. 5-11.** Major Elements of Call Store

processors are installed, each pair of signal processors can have up to eight call stores.

Call store functions may be divided into four main categories: memory module, associated memory circuits, logic and sequence circuits, and maintenance circuits. Figure 5-11 diagrams these main functions.

There are four memory modules in a call store. Each module has 2,048 binary word locations or a total of 8,192 binary words in a call store. The memory modules contain the ferrite sheet cores as later explained. They are divided into two information blocks designated the H half and the G half. This division is similar to that created in program store.

Referring to Fig. 5-11, the input bus selection circuit connects call store to one of the duplicated buses from central control. The output bus selection circuit similarly selects one of the output or answer buses used to send the readout binary word to central control. The memory circuits, which surround the four memory modules, control the access, inhibit, and readout operations. The access circuits are directed to the desired binary word location in a particular memory module by a 12-bit address word from the address register. The selection of the specific H or G half is determined by the binary digit derived from the 6-bit name code that is sent to the logic and sequence circuits. The data register holds the 24-bit data pulses, which were received from central control, for subsequently writing a binary word into call store. This register also is used in the readout mode to hold the particular readout binary word until it is rewritten into the memory module.

The logic and sequence circuits decode the 2-bit readout or write order, the 6-bit name code, and the 3-bit mode order words. One bit, as previously mentioned, is derived from the name code and used to select the H or G half of the memory module. The 3 bits in the mode order determine whether or not the particular call store should respond and the mode of operation, that is, whether a binary word is to be read out from or written into the memory module.

The maintenance circuits check the validity of the single parity pulse generated by the name code and the address. These circuits also verify test points within call store. If these test points and the parity pulse are found to be correct, a single bit pulse, designated "All Seems Well" (ASW), is generated and sent back to central control over the output bus.

## Call Store Memory Cell

The ferrite sheet with its 256 holes or cores is the basic storage element in the memory modules of a call store. Each ferrite sheet is approximately one-inch square and about 0.030 of an inch thick. The sheets are arranged in three adjacent stacks of sixty-four sheets each, or a total of 192 ferrite sheets in one memory module. A single ferrite sheet contains 256 holes, each 0.025 of an inch in diameter and placed on 0.050-inch centers. These holes are formed into a 16 by 16 grid or matrix with each hole providing storage for one binary digit. They may be considered as equivalent to an array of 256 miniature magnetic cores. Figure 5-12 illustrates a typical ferrite sheet used in the memory module.

The equivalent cores in the ferrite sheet have square-loop magnetic characteristics similar to the ferrite cores in program store which were explained previously and shown in Fig. 5-1. One binary digit can be stored in the ferrite material surrounding each hole by magnetizing it in either a clockwise or counterclockwise direction. A clockwise magnetization will store a binary 0, and a binary 1 will be stored by a counterclockwise direction of magnetization.

The holes in the ferrite sheet, as previously explained, correspond to a ferrite core memory cell as used in program store. Four conductors are associated with each hole in the ferrite sheet which may be considered as a memory cell (see Fig. 5-13). They are the X and Y selection wires, the inhibit wire,

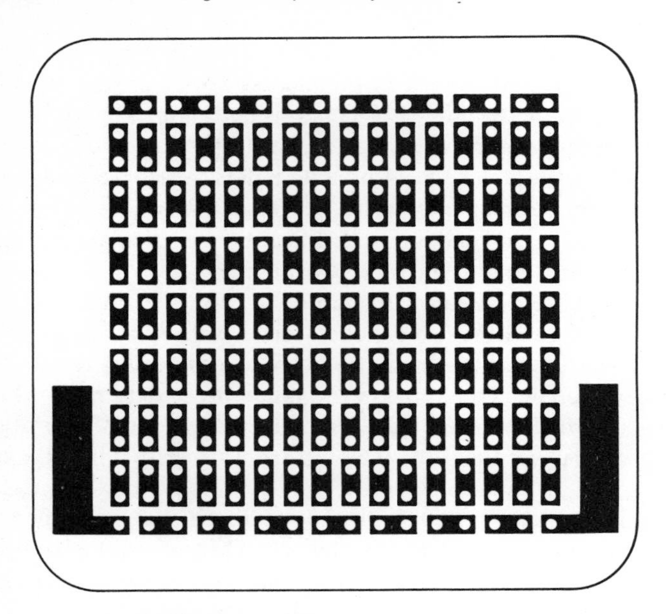

**Fig. 5-12.** Ferrite Sheet Used in Memory Module

and a readout wire. The Y select wire actually is a printed conductor on both sides of the ferrite sheet. The printed wiring also forms a metallic sleeve through the hole to connect together the printed conductors on the opposite sides of the ferrite sheet. Each row of holes is connected to the next row so that all holes in a ferrite sheet connect to the same Y select wire. The X and Y selection wires carry the dc pulses to access the desired hole or memory cell in a similar manner as described for the ferrite core in program store. Pulses representing binary digits read out from the memory cell are induced in the readout or sense wire. The inhibit wire is utilized to prevent a memory cell from switching or changing its magnetization polarity during the writing operation. For this purpose, a dc pulse is sent over the inhibit wire simultaneously with the current pulses over the X and Y select wires, but in the opposite direction.

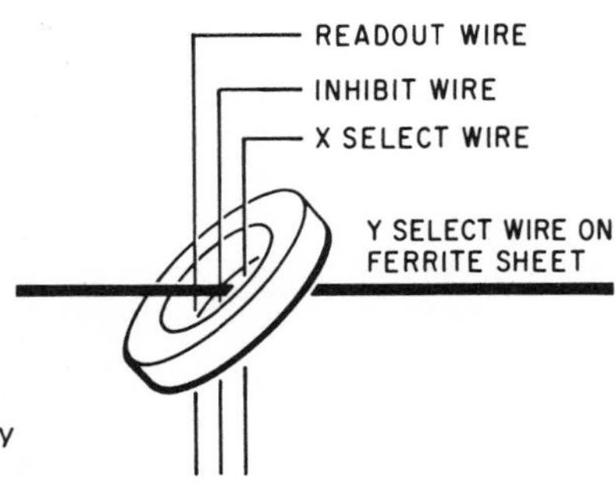


**Fig. 5-13.** Equivalent Ferrite Memory Core Used in Call Store

Each hole in the ferrite sheet, as previously mentioned, can store one bit in its surrounding material. To handle the 24 bits comprising a binary word in each call store's module, the sixteen rows of holes in the ferrite sheet are subdivided into eight row pairs. This results in 8 bit levels or planes. The readout or sensing wire is coupled through two adjacent rows for each bit. Thus, by utilizing three stacks of bit pairs, the required 24 bits in the binary word will be collated.

## Readout Operations in Call Store

Readout and writing operations in call store are carried out in two stages. The first stage in the readout operation is to gain access to the location of the required binary word in the memory module. This is accomplished by sending a current pulse simultaneously over one of the sixty-four X selection wires and one of the sixty-four Y selection wires in the specified H or G half of the particular call store. These current pulses are bipolar, that is, a negative current pulse is first sent and then immediately followed by a positive pulse. The negative pulse is used to read out the memory cell which is at the intersection of the specified X and Y selection wires.

The combined current pulses over the X and Y leads will reset the particular memory cell, that is, its magnetic state will be switched to that which is associated with binary 0. If the memory cell already is in the binary 0 state, there will be a negligible change in magnetization. Consequently, a very low output voltage will be induced in the readout or sensing wire. Now, if the memory cell's initial magnetic condition should correspond to the binary 1 state, the aforementioned combined X and Y current pulses will cause a change in its magnetization. This magnetic change will induce an appreciable voltage in the readout wire. Thus, no voltage in the readout wire indicates that the memory cell initially was in the binary 0 state, while the presence of a voltage indicates an initial binary 1 state. The resultant binary digit outputs are stored in the data register illustrated in Fig. 5-11.

Irrespective of its initial state, the memory cell is always changed into the binary 0 state by the readout operation. Consequently, the memory cell in call store is considered to have a destructive readout. It is essential that the information previously stored in the memory cell not be lost. Therefore, during the second stage of the readout operation, the circuits in the memory module are designed to write back whatever was read out.

For example, to write back a digit, the location of the previously readout binary word again is selected by pulsing the same X and Y selection leads. But the direction of the current pulses now is opposite to that used for the readout operation. Therefore, each selected memory cell will be set, that is, switched from the binary 0 state to the binary 1 state. Consequently, the X and Y current pulses would tend to write back a binary 1 in each bit of the selected binary word location. However, any bit in the data register that is equal to 0 will cause an associated inhibit circuit to function. This circuit will apply a negative current pulse to the related inhibit wire. The resultant magnetic field, due to the inhibit wire and the opposing magnetic flux of the X and Y selection wires, will not be sufficient to switch the memory cell. Accordingly, the memory cell's magnetization is not changed and it will remain in the binary 0 state.

## Writing Operations in Call Store

To write new information in the form of binary words into a memory module of call store, a readout operation is first performed on the memory cells of the binary word in the specified location of the H or G half of call store. This is necessary so that all 24 bits of the selected binary word will be switched to the binary 0 state if any should be in the binary 1 state. The bits which are read out, however, are not stored in the data register as in the case of a normal readout operation.

The effect of the first stage of a writing operation, therefore, is to clear or reset to zero the binary word in the specified location. Next, the information to be written is received over the input bus from central control, in the form of a 24-bit binary word, for storage in the data register. During the second stage, this information is taken from the data register. It is written into the selected binary word location on the ferrite sheets of a particular memory module in the manner previously described.

For example, to write a binary 1, positive pulses are applied to the appropriate X and Y selection wires to create a counterclockwise magnetic field in the selected memory cell. Memory cell A in Fig. 5-14 illustrates this condition. The effect of this action is to reverse the magnetization of the cell which

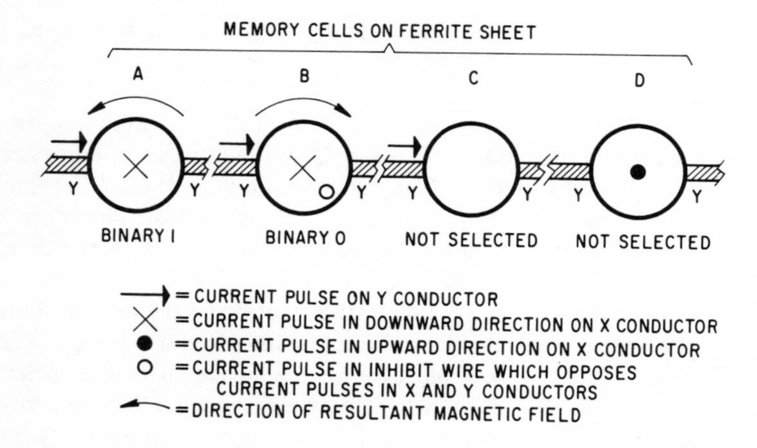


**Fig. 5-14.** Writing Binary Digits in Memory Cells of Call Store

was left in the binary 0 state by the readout operations during the first stage. As a result, a binary 1 will be stored in this particular cell.

A binary 0 is written into a memory cell by not changing its magnetic state from that resulting from the readout operations during the first stage. To achieve this, a negative current pulse is applied to the inhibit wire at the same time that the positive current pulses are sent through the X and Y selection wires. Memory cell B in Fig. 5-14 indicates this condition. The resultant magnetic field will not be sufficient to switch this memory cell's magnetization from a clockwise direction and it will remain in the binary 0 state. Memory cells that are not selected by coincident current pulses in the X and Y conductors retain a weak remanent magnetic field. Memory cell C in Fig. 5-14 illustrates this situation for a current pulse only on the Y conductor of an unselected memory cell. Memory cell D represents a current pulse in the upward direction on the X wire only of a similar unselected cell.

## Memory Cell Access Circuits in Call Store

A particular memory cell on a ferrite sheet in a module of call store is selected by its associated X and Y access circuits. Figure 5-15 is a schematic drawing of the selection arrangements. The access circuits are controlled by the 12-bit address information received from central control and recorded in the address register as shown in Fig. 5-11. The first six of the 12 address bits control the Y access circuits. The other six address bits similarly control the X access circuits.

Two X and two Y access circuits are provided to select the desired memory cell in a memory module. One X and one Y access circuit are associated with the H half of the module and, likewise, one X and one Y access circuit are utilized with the G half. Thus, an access circuit can select one out of sixty-four memory cells on the ferrite sheets in the desired H or G half of the memory module.

The X and Y access circuits are connected through a number of electronic switches to dc pulse generators or drivers as shown in Fig. 5-15. Separate readout and write current generators are provided to pulse the X and Y selection leads, first in one direction and then in the opposite. Each electronic switch consists of solid-state devices (transistors and diodes) which allow current to flow only when a particular combination of address bits is present. Each electronic switch requires a different combination of three address bits for operation.

For instance, referring to Fig. 5-15, during a readout operation, current can flow through only one "A" switch at a time to the associated X access circuit whenever the specified combination of three address bits (in the A6 to A11 sequence) is received. The desired "B" switch is controlled in a similar fashion to connect to the X access circuits in connection with a writing oper-

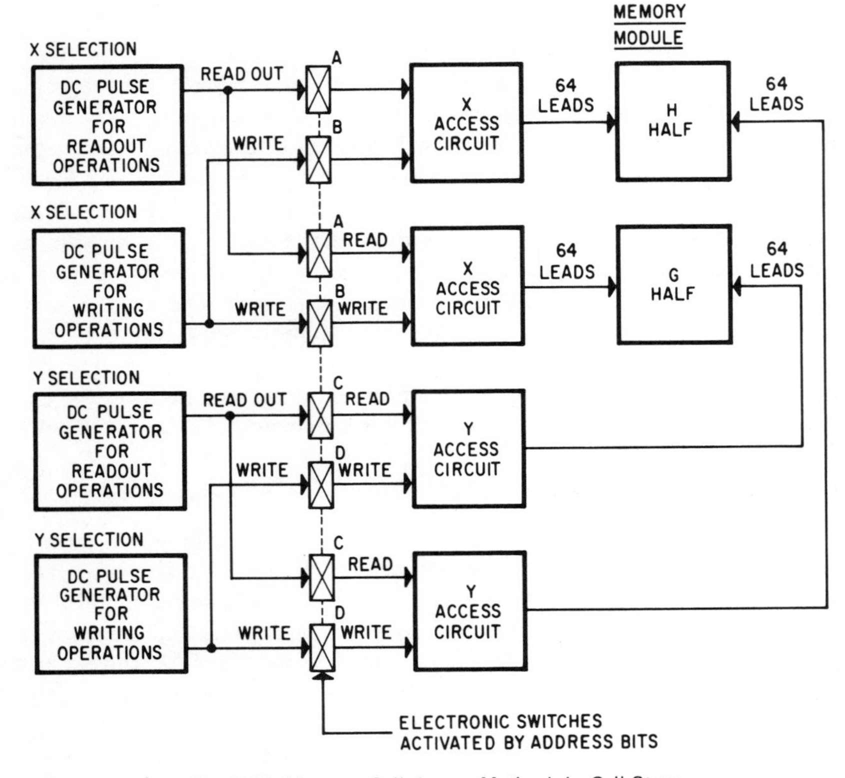


**Fig. 5-15.** Memory Cell Access Methods in Call Store

ation. The Y access circuits are associated with the "C" and "D" electronic switches for readout and writing operations, respectively. A combination of three address bits in the A0 to A5 sequence is utilized to control these switches which function in the same manner as described for the "A" and "B" electronic switches. In this way, a current pulse will flow through only one X wire out of sixty-four, and through only one Y conductor out of sixty-four in a particular memory module of call store. The same X and Y wires will be pulsed in the opposite direction during a writing operation.

## Duplicate Input and Output Operations of Call Stores

Requests from central control for readout or writing operations are accomplished simultaneously by two different call stores. The H and G halves of the information blocks in every call store are assigned a 6-bit name code as previously explained. The H half's code name is established by appropriate wiring at the time of installation. On the other hand, the G half is assigned a code name by the action of its flip-flop circuits under direction of central con-

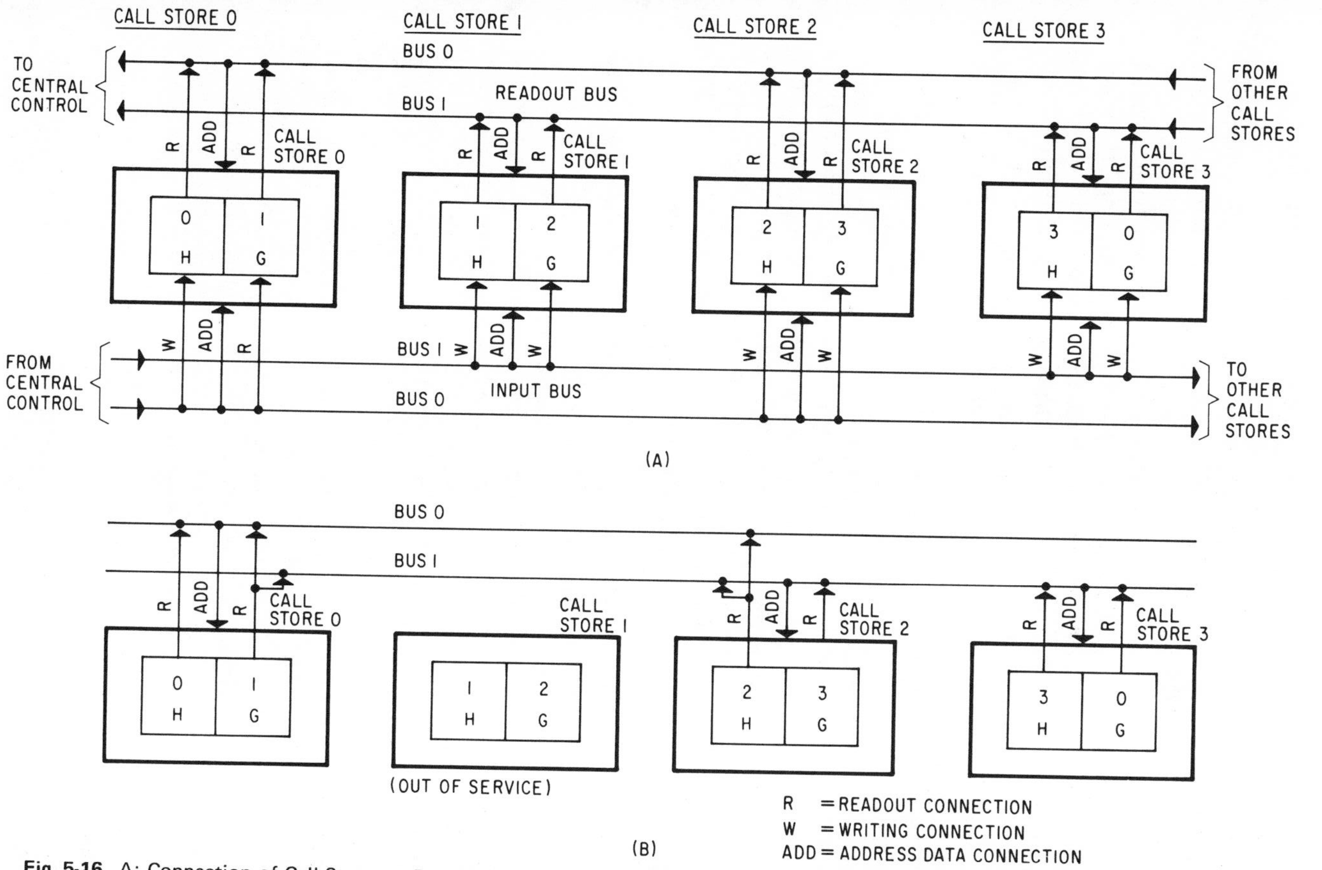


**Fig. 5-16.** A: Connection of Call Stores to Duplicate Buses B: Connection of Call Stores to Buses When Call Store 1 is Out of Service

trol. This coding procedure permits selected call stores to be connected to central control utilizing a common bus system.

Figure 5-16A illustrates a typical connection of four call stores, containing information blocks 0, 1, 2, and 3 to the duplicate input and output buses. For instance, assume that central control is requested by a program to read binary word A which is located in information block 2. Two copies of this binary word are available, one in the G half of call store 1 and the other in the H half of call store 2. Central control identifies the desired binary word by means of an 18-bit address code. This address code is divided into a 6-bit name code, K, which indicates the location as in information block 2, and a 12-bit address, A, which specifies the particular location of binary word A within block 2.

The name code, K, and the address, A, are transmitted by central control over both readout buses 0 and 1 to all call stores. However, only call stores 1 and 2 will detect a match between the K name code and their internally assigned name codes. Consequently, only these two call stores will utilize address A to read out the desired binary word. For example, the G half access circuit would be activated in call store 1 because its name code matches the K name code that is received. Similarly, the H half access circuit in call store 2 will be activated. Moreover, bus 1 transmits the address data and receives the readout from binary word A in the G half of call store 1. Bus 0 supplies the address data and receives the readout from the duplicate binary word A located in the H half of call store 2. The same procedure is followed for all other information blocks in call stores.

The duplicate buses and the control circuits within call stores provide means for overcoming trouble conditions that may be encountered in call stores or its interconnecting buses with central control. For instance, call store 1 has been taken out-of-service because of a trouble condition. The flip-flop circuits of call store 1, therefore, will reject any requests for its information from blocks 1 or 2. Consequently, only the G half of call store 0 will be available to handle requests from central control for information from block 1. The readouts from this G half, therefore, will be sent on both buses 0 and 1 as shown in Fig. 5-16B. For the same reason, only the H half of call store 2 will be able to handle requests for information from block 2. As a result, the readouts from this H half, likewise, are sent over both buses. This flexible arrangement for the call stores and their interconnecting buses is similar in many respects to that provided for program stores as previously explained.

## Questions

1. What two types and designations of memory systems are utilized by central control in processing calls? In what form are the instructions and information stored in these memory systems?

2. What is the basic element of the memory cell used in both memory systems? What are its main characteristics? What conductors intersect in the memory cell and what are their purposes?
3. What main function is performed by a memory cell in program store? In call store? What is the relationship between the direction of magnetization of a memory cell and its stored digit?
4. What major elements comprise the basic memory module of a program store? Describe the composition of the aluminum memory card.
5. How many memory modules are installed in one program store? What is the storage capacity of a program store in binary words and bits? How are the binary words divided in a program store? Why?
6. During the readout process of a particular aluminum card in program store, a specific bar magnet on the card is not magnetized. What is the binary digit readout? What would be the readout if this bar magnet were magnetized? Why? What equipment is utilized to change translation and related data in the aluminum memory cards of a memory module?
7. What operations are performed by call store? What three or more types of information are stored in the call store memory?
8. How many binary word locations are contained in each memory module of a call store? What is the total binary word capacity in one call store? How are these binary words divided? How many bits are contained in each binary word of call store?
9. What is the basic storage element in the memory module of call store? What function does it perform? What conductors pass through this element? Briefly describe their functions.
10. What happens to the stored information in a memory cell during the readout operation in call store? Briefly describe the writing operation in call store.

# 6
# STORED PROGRAM CONTROL AND CALL HANDLING OPERATIONS

## Stored Program Organization

The stored program in the No. 1 ESS and similar electronic switching systems is designed to provide real-time operations of the switching system in response to random traffic, or varying customer calling rates. For instance, a customer expects to hear a dial tone immediately upon picking up his handset. Likewise, upon completion of dialing the customer expects to hear the ringing signal within a second or two. Random traffic refers to the fact that calls may be handled at an average rate of about six per second in the busy hour. During one second, however, no calls or only two or three may be processed, while during another second as many as twelve may be initiated.

The electronic switching system, consequently, must be capable of processing this random traffic as quickly as possible in order to approach the real-time requirements for good service. To fulfill this requirement is a main function of the stored program control system in the No. 1 ESS. It is designed to handle as many as 100,000 calls in the average busy hour and also has the capability of handling new service features as well as future growth requirements. In addition, the stored program control arrangement provides for automatic detection and diagnosis of trouble conditions, and for use with a signal processor if provided or added later.

The stored program part consists of instructions, translation information and related data that are stored in program store. It also includes the latest changes concerning customer lines and trunks that are in the temporary memory of call store. Figure 6-1 portrays the main elements of the stored program control system. In essence, the stored program consists of approximately 100,000 instructions of which a little over half are devoted to call processing and related functions, with the balance covering automatic maintenance operations of the electronic central office. These instructions are read by central control one at a time from program store or from call store as needed. Note that neither the stored program nor central control by itself can

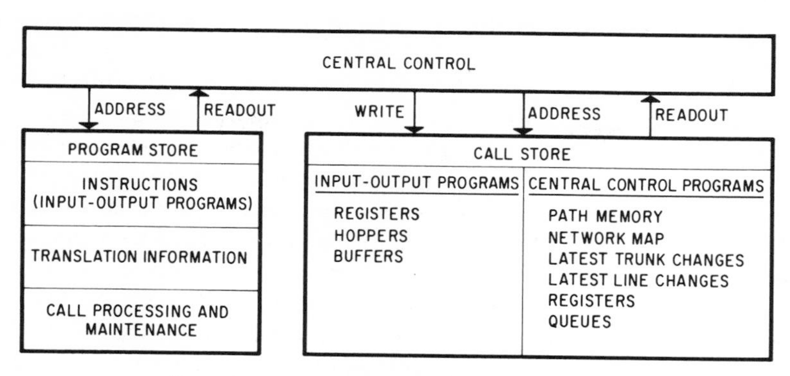


**Fig. 6-1.** Stored Program Control System Elements

carry out any functions. Rather, central control is the means of implementing the actions ordered by the stored program. For example, a function such as detecting a customer's line request for dial tone results from the combined effects of the program instructions read one at a time from program store and executed by central control.

## Stored Program Control Analogy

In Chapter 2, the operations of electronic switching systems were explained by an analogy to the operations of the antiquated manual switchboard which was assumed to be run by electronic operators or automatons instead of human operators. This same analogy will be used to simplify the explanations of the operations of the stored program control system and its program organization. For this purpose, we will assume that the switchboard is controlled by an electronic supervisor and an electronic operator.

The work performed by the electronic operator is not complicated but it is time consuming. It includes scanning the face of the switchboard for lamp signals indicating calls being originated, watching the cord lamps for a lighted lamp indicating disconnection of a call, obtaining the called number from the calling customer, and plugging and unplugging switchboard cords. These tasks may be classified as input-output actions or work performed in accordance with input-output programs. It is essential that these tasks be carried out at regular intervals to avoid loss of information and resultant downgrading of service.

The electronic supervisor's tasks are more complicated because they involve actions such as analyzing the significance of digits received for various calls in progress, deciding what connections must be established or taken down at the switchboard, and recording the necessary information needed for billing of calls. These particular tasks, however, can be deferred for a short interval without affecting service.

To record the aforementioned types of information, the electronic supervisor and electronic operator use separate electronic slates. The electronic slate used by the electronic supervisor corresponds to the call stores that are directly associated with central control in electronic offices where signal processors are installed. Figures 2-5 and 4-4 illustrate the role of signal processors. The electronic operator's slate is analogous to call stores associated with signal processors. These electronic supervisor-operator relationships are depicted in Fig. 6-2. In electronic central offices not equipped with signal processors, the electronic supervisor would handle both its job and that of the electronic operator. In this case, separate areas of the same electronic slate (call store) are provided for each of the previously mentioned tasks.

The electronic operator communicates with the electronic supervisor by recording information in certain areas of its electronic slate called *hoppers*. For example, when an originating call is detected, the electronic operator enters this information in a list designated *service request hopper.* At a later interval, the electronic supervisor will unload or take out, one at a time, the data in this hopper in order to carry out the required action for each entry.

The electronic supervisor communicates with the electronic operator by recording data in areas of the latter's slate known as *output buffers.* For in-

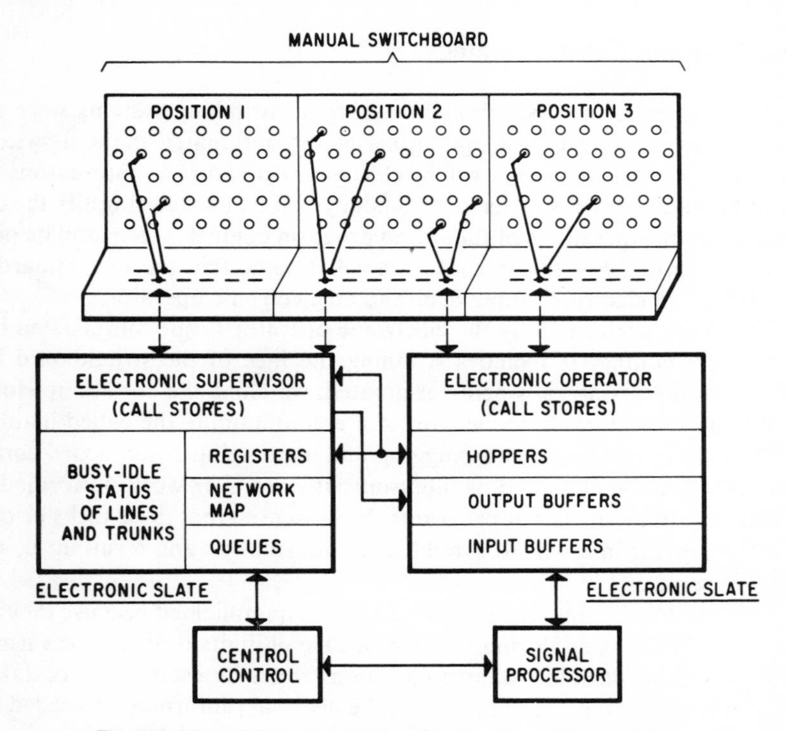


**Fig. 6-2.** Manual Switchboard Analogy for Stored Program Control

stance, when the electronic supervisor has decided what particular switchboard cords will be utilized to plug into specific jacks in handling a call, it will write the appropriate information into an available output buffer in the operator's electronic slate. The electronic operator, will subsequently unload this information from the output buffer and will then proceed to perform the required actions, such as inserting cords into specified jacks on the switchboard.

Other portions of the electronic supervisor's slate are used to record different types of information. For example, the busy-idle status of all switchboard cords (corresponding to the line-link and trunk-link network) is recorded in a certain part of the electronic slate (analogous to the network map in call stores). Other areas, called ***registers***, are used to record data pertaining to the progress of calls. In this connection, the electronic supervisor selects and assigns a register to each call originated by a customer. This register is used by the electronic supervisor to accumulate the digits of the called number. These digits are received from the electronic operator via the digit hopper in which they are initially recorded, one at a time, as submitted by the calling customer.

In the course of handling a call, the electronic supervisor may find that certain facilities, such as a register, are not available because of traffic loads. In this event, it will record this information in an area of the electronic slate designated as the ***queue*** or waiting list for registers. This list or queue will be frequently checked by the electronic supervisor to ascertain whether it is possible to obtain a register to serve the waiting customer lines. The queue, in effect, serves as a memorandum to remind the electronic supervisor about tasks waiting to be accomplished.

## Call Processing Analogy

When signal processors are not provided, as in small and medium-size electronic offices, the electronic supervisor in our analogy also performs the electronic operator's tasks as previously stated. The electronic supervisor follows a prescribed schedule in carrying out the various call processing tasks. For instance, a particular task is initiated by checking a hopper or queue for any recorded data. Next, actions are performed which may involve recording information in a register or in an output buffer. The electronic supervisor proceeds from one item or entry to the next one until the bottom of the hopper or queue is reached. Then, it continues to the next hopper or queue in accordance with the list of tasks to be accomplished. When the end of the list is reached, the electronic supervisor starts over again.

Every 5 milliseconds, a clock circuit will signal the electronic supervisor to interrupt work and to record the data on the electronic slate. In this manner, it will be able to resume exactly where it left off when the clock signal was received. The electronic supervisor now undertakes a scheduled list of jobs which would have been accomplished by an assisting electronic operator if one were provided. Some of these tasks are performed during the interrupt signal every 5 milliseconds, while others are carried out every 10 or 25 milliseconds. The amount of time spent by the electronic supervisor on these secondary tasks depends upon the traffic load. Normally, the electronic supervisor would resume the main tasks after about 1.5 milliseconds.

Consequently, when a signal processor and its associated call stores are provided, the electronic supervisor can handle a greater volume of calls. For example, the electronic supervisor can continuously attend to the main functions of unloading hoppers and loading output buffers without interruptions. At the same time, the assisting electronic operator (utilizing the signal processor and its associated call stores) can follow the usual 5-millisecond schedule of loading hoppers and unloading output buffers.

## Stored Program Functional Groups

The stored program, for purposes of call processing, consists of the following four functional groups with each group controlling a particular stage in the call processing procedure:

1. *Input-Output (I/O) Programs.* The input programs gather particular types of information for call processing purposes. For example: customer's dial pulses and digits received, line disconnections, and the idle or busy state of all lines, trunks, and service circuits. The output programs transmit data, obtained from the processing operations, to the switching network and other peripheral equipments. This output program data results in the operation and release of ferreed switches that provide interconnections in the switching network, and in the operation of relays in trunk and service circuits.
2. *Operational or Call Processing Programs.* These programs are concerned with examining and processing the information pertaining to a particular type of call or to a phase of a call. These programs can also call on subroutines to collect data for use by the output programs. For example, the translation subroutine may be requested to supply data for the dial connection program.
3. *Service Routines.* Programs that process functions which are not solely related to one type of call or to a particular aspect of a call are known as service routines. An example is a program requesting a translation service routine to determine the specific line scanner equipment number that corresponds to a given directory number.
4. *Executive Control Program.* This is the main program which schedules the work of the input-output and call processing programs.

Each of the various programs in the stored program control system assumes and relinquishes direction of central control on a strict time schedule that is governed by the executive control program.

## Stored Program Functions of Call Store

Call store is directly associated with the stored program control system because its temporary memory acts as a clearing house for information between the above functional groups. Call store, for this purpose, is divided into sections or areas, each with a number of binary words or memory slots. These areas have registers, hoppers, and buffers in which information is deposited and withdrawn as required in processing a call.

For instance, one or more registers is associated with each call being processed. The hoppers receive information from the input programs and this information is then processed by the operational programs. In turn, the operational programs together with the subroutines send the resultant data to the buffers for use by the output programs. Moreover, the information in many areas of call store is constantly modified or updated as a call is processed. Data recorded by one program may again be used later by the same or other programs. Thus, the handling of any call involves a constant interaction between program store and call store with central control carrying out the resultant orders.

Each area in call store consists of one or more binary words of 23 bits each. All of the 23 bits may be utilized to store a particular information item or the binary word may be divided into parts of one or more bits. For example, the entire 23 bits of a word may be used to store the identity of the trunk-link network termination of a service circuit connected for a particular call. Groups of 4 bits may be used to store the three digits of the called office code. Likewise, groups of 4 bits in another binary word may be utilized to record the called line number. Figure 6-3 illustrates this use of the call store area. The first and second binary words are used to store the called number

BIT POSITION NUMBER

23 — 15 — 10 — 5 — 0

| BINARY WORD | 23 | GROUP 6 | GROUP 5 | GROUP 4 | GROUP 3 | GROUP 2 | GROUP 1 | DECIMAL NUMBER |
|---|---|---|---|---|---|---|---|---|
| FIRST | 1 | 0 0 0 | 0 0 0 0 | 0 0 0 0 | 0 1 0 1 | 0 1 0 1 | 0 1 0 1 | 555 |
| SECOND | 1 | 0 0 0 | 0 0 0 0 | 0 0 0 1 | 0 0 1 0 | 0 0 0 1 | 1 0 1 0 | 1210 |
| THIRD | 1 | 0 0 0 | 0 0 0 0 | 0 0 0 0 | 0 0 1 1 | 0 1 0 0 | 0 0 1 0 | 342 |
| LAST BINARY WORD | 1 | 0 1 0 | 1 0 1 0 | 0 1 1 0 | 0 1 1 1 | 1 0 1 0 | 0 0 0 1 | 206701 |

BIT NO. 23 IS USED FOR PARITY CHECK PURPOSES

**Fig. 6-3.** Example of Stored Binary Digits in Call Store Memory Area

by its office code and line digits of 555–1210. The third binary word identifies the trunk-link network termination as 342. The equivalent decimal numbers of the binary digits in the 4-bit groups are derived from the binary coded decimal table illustrated in Table 3-1.

## Program Priorities

Since call processing operates in real time, the stored program system must respond immediately to signals and data that occur at times that are not under its control. Central control, for instance, must keep up with the flow of information from customers and other central offices. To ensure dependable operations, it must respond quickly to errors detected by the many trouble detector circuits in the various equipment units. Otherwise, improper handling of calls and poor service may result. Consequently, it is necessary to establish an order of priorities in the system to ensure that operations of a *nondeferrable* type are carried out under controlled time schedules.

For example, detecting dial pulses is a nondeferrable task because pulses may be missed if this program is postponed for 5 milliseconds. Deferrable programs are those that can be deferred for 5 or more milliseconds without noticeable effects on service. Maintenance tasks may be either deferrable or nondeferrable depending upon the particular task. That is, if a call store unit becomes faulty, it is very important that it be immediately identified and removed from service. The diagnosis of a junctor link that has been removed from service, however, is a deferrable job that can be accomplished at other times.

A program being executed may be immediately stopped by an interrupt signal generated from one of several sources. In this case, a transfer is usually made to another program associated with the source of the interrupt signal. When the interrupt program completes its functions, the particular program that was interrupted resumes its functions as if it had not been interrupted. The interrupt signals enable central control to accomplish input-output operations that must be carried out at regular intervals, such as detecting dial pulses, scanning for line disconnections, and detecting ringing and answer signals. In addition, the interrupt signals cause central control to immediately take appropriate actions when trouble conditions are discovered, or if the execution of certain programs should be requested by maintenance personnel.

## Program Interrupt Levels

Program interrupt operations may be divided into two major categories: interrupts due to the system clock which occur every 5 milliseconds, and maintenance interrupts which may happen any time as the result of trouble detection. Program interrupts caused by maintenance have higher priority than those due to clock interrupts. For instance, if a maintenance interrupt occurs when central control is under the direction of the executive control program, central control will be transferred to the proper fault recognition program. In this manner, the defective unit can be identified and taken out of service. At the same time, the required information is entered in the maintenance register to permit the necessary diagnostic work to be accomplished later on a deferrable basis. The call processing work is then resumed. The interruption period in this case is normally less than 40 milliseconds.

The interrupt sources and the programs associated with them are classified into ten levels of importance. These interrupt levels are designated from the highest to the lowest order by the letters A to K, inclusive, except for the letter I. An interrupt source allotted to a specific level can only interrupt programs of lower levels, except that level B can interrupt level A. Most of the call processing programs in the No. 1 ESS are not related with any interrupt source. Consequently, they may be interrupted by any order of interrupt level. These programs are considered as operating at the lowest or base level L. Table 6-1 illustrates the arrangement for these program interrupt levels.

The A level, which is the highest interrupt source, is initiated from the *master control center*. Levels B through G are interrupted by trouble detectors in the electronic switching equipment. Interrupt levels H and J are activated every 5 milliseconds when a signal processor is not installed. The J level initiates the input-output programs associated with central control. The H level is used to interrupt the J level's input-output programs whenever more than 5 milliseconds are required for their processing tasks. Interrupts of level K are generated only in electronic offices equipped with signal processors when, for instance, a hopper overflows because of traffic loads. Base level L constitutes the *task dispenser programs* which are used to control specific hoppers or queues in call stores. Functions of the task dispenser programs are scheduled by the executive control program which is the main program.

**Table 6-1.** Priority Rank of Program Interrupt Levels

| Interrupt Level | | Interrupt Source |
|---|---|---|
| Maintenance Programs | A | Master control center |
| | B | Emergency action |
| | C | Central control |
| | D | Call store |
| | E | Program store |
| | F | Peripheral units |
| | G | Special tests |
| | H | Input-Output programs* |
| | J | Input-Output programs* |
| | K | Hopper overflow (used only when signal processors are installed) |
| | L | Base-level programs (directed by the executive control program) |

* Interruptions of H and J levels occur every 5 milliseconds

## Executive Control Program

The executive control program schedules the work of the other functional program groups. For instance, every 200 milliseconds, the executive control program schedules the operation of the line scanning program in the input program group. This particular program directs central control to scan all customer line ferrods to ascertain which have changed state since the previous scan. Whenever a change is detected, the line scanning program temporarily stops and the equipment number of the calling line is written into the line service request hopper in call store. The required action on one line is completed before proceeding to scan the next line's ferrod. When all customer line ferrods have been scanned, the line scanning program returns control to the executive control program. At this time, any customer's line that has been off-hook since the last scan will be identified in the line service request hopper. These program functions are represented by the applicable block diagrams in Fig. 6-4.

The subsequent action to be taken is next decided by the executive control program. For instance, it could schedule an operational program, such as the dial connection program, to handle dial tone requests that are stored in the line service request hopper in call store. It may decide, instead, to allocate control to the line link switching network or to another output program. In

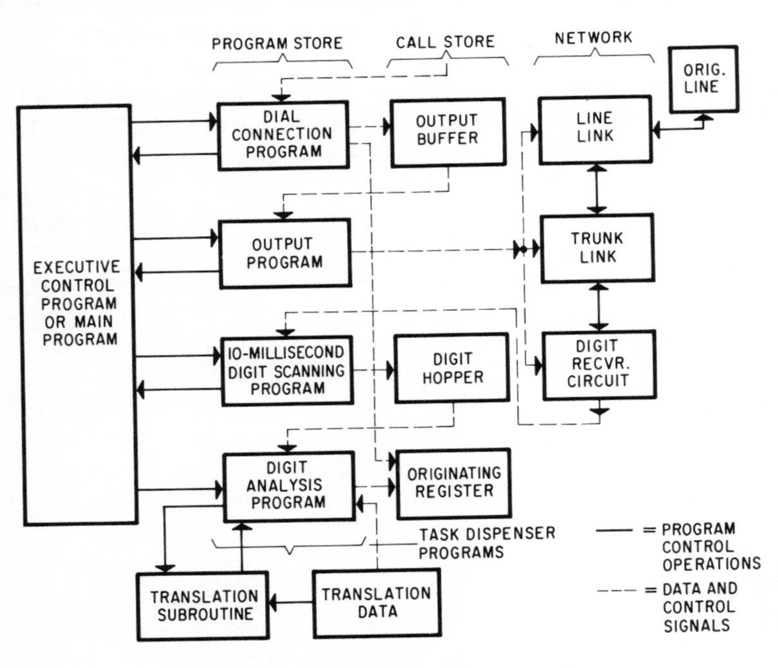


**Fig. 6-4.** Example of Task Dispenser Programs

this case, a particular controller in the line-link network can operate only once in every 25 milliseconds. Therefore, the executive control or main program would direct other system operations in the interval.

Meanwhile, the service requests in the line service request hopper of call store are waiting for an operational program. Consequently, one step in processing a call may have no direct relation in time with the next step. Thus, as a specific call progresses through the electronic central office, control of its operation is handed back and forth between the main program and the operational programs within program store. Moreover, an operational program, for example, the ringing trip scanning program, may complete a processing action on one customer's line and then return control to the executive control program. Thus, the main program could direct the same operational program to the next waiting customer's line or it may initiate another operational program. In any event, a particular operational program is interrupted whenever a digit scanning program or network operation step is due.

## Task Dispenser Programs

The executive control program also schedules the operations of task dispenser programs. A task dispenser program takes data that was entered in a hopper or queue within call store, and refers it to an appropriate task program in program store for necessary processing. Consequently, task dispenser programs serve as links between the main program and the various operational or task programs that carry out the processing actions on the data in the buffers, hoppers, and queues in call store. Figure 6-4 shows a number of task dispenser programs that are utilized in program store. A brief synopsis of the operations involved follows.

Assume that a service request (customer's line off-hook) is received by the dial connection program from the line service request hopper (see Fig. 6-5). The executive control program first directs the dial connection program to load or temporarily store this request in an available output buffer in call store. Therefore, when scheduled by the main program, the output program will cause the output buffer to operate the appropriate controllers in the line link and trunk link networks. These actions will connect an idle dial pulse or digit receiving circuit to the originating customer's line. A dial tone will be sent at this time.

Every 10 milliseconds, all dial pulse or digit receiver circuits are scanned under control of the digit scanning program. All detected dial pulses are initially entered in the digit hopper in call store. The digit analysis program, as soon as scheduled by the main program, will direct the digit hopper to enter the recorded digits into an originating register in call store. Then, the digit analysis program with the aid of its translation subroutine and the translation data in program store will proceed to further process the call.

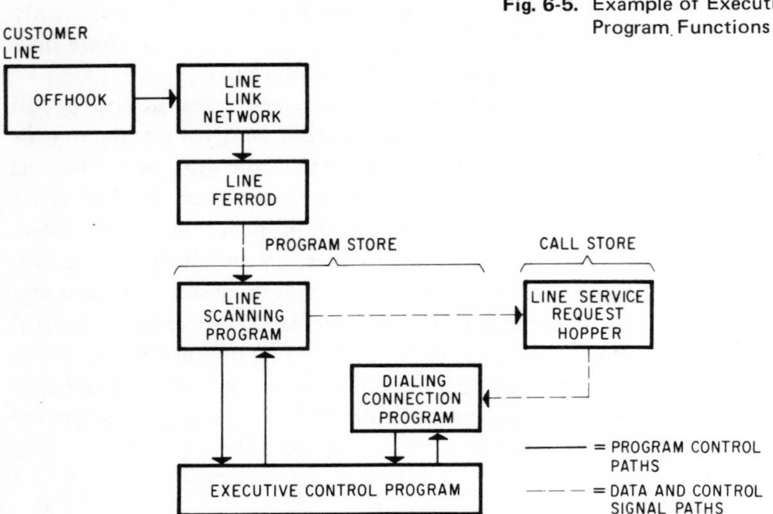


**Fig. 6-5.** Example of Executive Control Program Functions

## Task Priority Classes and Interject Programs

The time required between the completion of a particular task dispenser program and its resumption depends upon the duration of all the intervening tasks. The buffers in call store, for instance, accumulate information of different degrees of importance and, consequently, must be unloaded in accordance with a precise schedule. The task dispenser programs, therefore, are grouped into five time priority categories designated A, B, C, D, and E, and also an "interject" class which has the highest priority. Each priority class has a class control program that directs the operation of the task dispensers in that class.

Class A task dispenser programs, for example, cannot be delayed more than 200 milliseconds. Class E, the lowest priority, however, may be deferred as long as 3 seconds. Thus, the executive control or main program is arranged to utilize a frequency table, consisting of a continuously repeating sequence of the five classes, to determine the next priority class to be activated. Each priority class in the frequency table is assigned twice as often as the priority class below it:

A B A C A B A D A B A C A B A E A B A C A B A D A B

Note that the sequence contains Class A thirteen times, Class B seven times, Class C three times, Class D twice, and Class E once.

The interject class is not included in the above sequence, but interject programs may interrupt at any point. The executive control program makes a check for the existence of any interject requests in each task program. Such requests may be initiated by programs in interrupt levels H, J, or L as shown in Table 6-1. The interject requests are originated whenever a program in the interrupt levels encounters a task that cannot tolerate the delay of even a Class A priority (200 milliseconds). In such a circumstance, the executive control program interjects the particular task that must be performed between other task programs.

## Translation Programs, Routines, and Subroutines

The translation programs in program store are used to convert known data, as supplied by a task dispensing program, into related data that is referred back to the requesting task program. The known data may be a three-digit office code, a four-digit directory number, or the identity of a trunk group or a network termination. A translation program that is common to a number of task programs is usually designated a *translation routine.* When a particular translation program or routine always returns control to its task program upon completion of its work, it is called a *translation subroutine.*

An example of the use of translation programs is illustrated in Fig. 6-4. Assume that the digit analysis program receives office code 393 from the digit hopper in call store. The digit analysis program will use this known data, 393, as an address to gain access to the related information in the translation subroutine. Translation data in program store are grouped into tables according to the information used to obtain access to them. Thus, the related translation information received by the digit analysis program for office code 393, would be loaded into an assigned originating register for processing.

Changes in translation data are sometimes required because of equipment reassignments or changes in service resulting from customer requests. Such modifications are not made immediately in program store, but are entered in the call store memory and designated as *recent changes.* Subsequently, these recent changes are collected and then written on the aluminum memory cards in program store by means of the memory card writer as explained in Chapter 5. Translation service routines usually consult the recent changes entered in call store before referring to the translation data in program store.

## Service Routines

During call processing operations, it is often necessary to initiate changes in the switching network and in the state of junctor, service, or trunk circuits. To accomplish these jobs, the task program calls upon appropriate service routines. For instance, the ringing connection program may request a network service routine to search for an idle ringing circuit. It may also request paths through the line-link and trunk-link networks to connect the ringing circuit to the called customer's line.

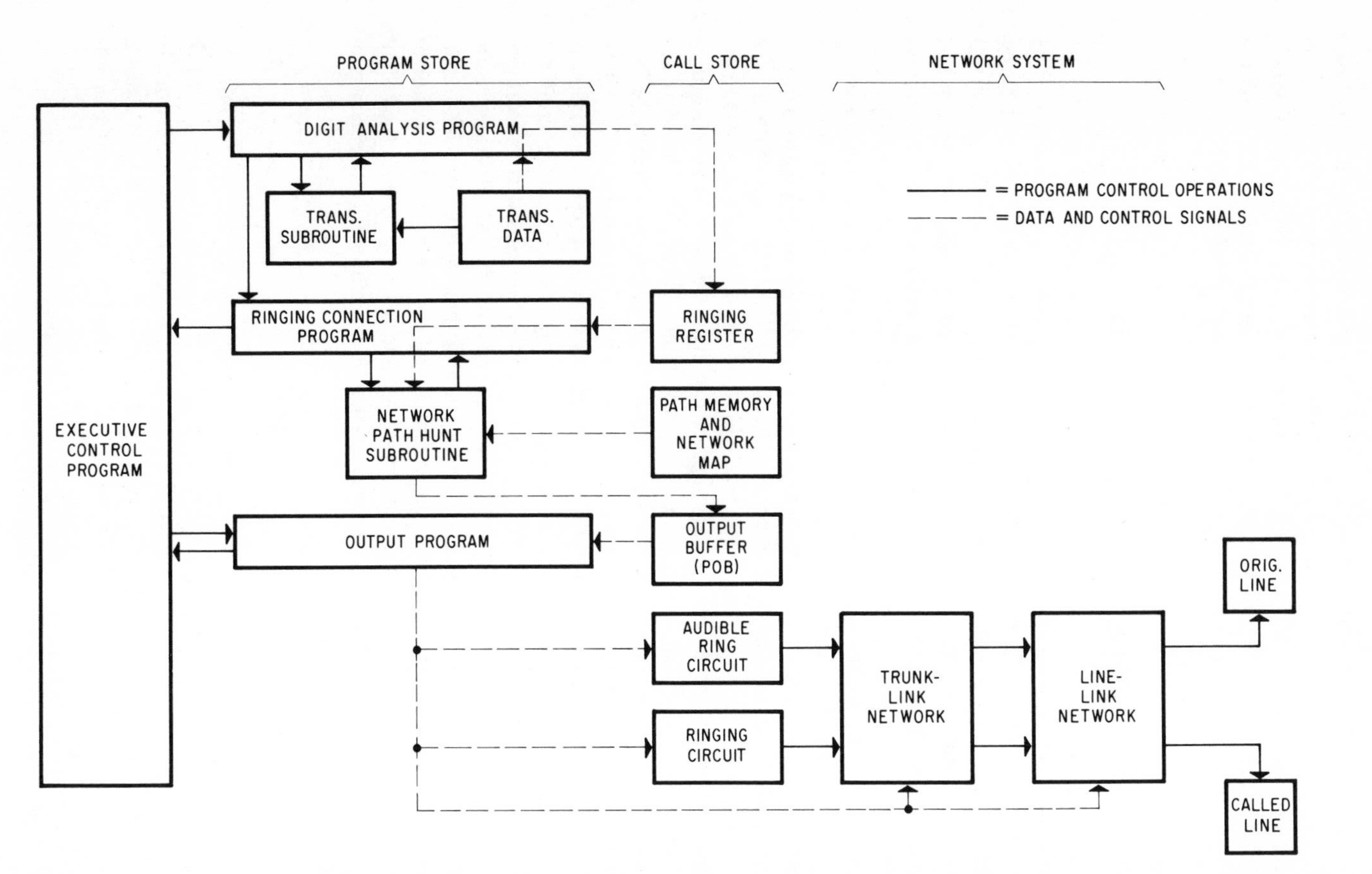


**Fig. 6-6.** Example of Translation and Network Service Routine Program Functions

An example of some functions of a network service routine during an intraoffice call is shown in Fig. 6-6. The ringing connection program, in this example, calls upon a particular network service routine designated the network path hunt subroutine. It requests this service routine for the related data needed to select a ringing circuit for connection to the called customer's line, and for an audible ring circuit that will be connected to the originating customer's line. The network path hunt subroutine first refers to the path memory and network map in call store for the related data on available audible ring and ringing circuits, and for the idle paths needed to connect these circuits to the network terminations involved. The selected paths are marked busy in the path memory and network map in call store.

The network service routine next loads an output buffer in call store, usually a peripheral order buffer (POB), with the necessary information for performing the required interconnections. The output buffer, as soon as activated by its associated task dispenser output program, will send the necessary control signals to make the required interconnections in the line-link and trunk-link networks as indicated in Fig. 6-6. Other service routines are used in a similar manner to place orders in peripheral order buffers (POB) for operating scanners, signal distributors, and central pulse distributors in connection with call processing.

## Network Map

A network map (see Fig. 6-7) is associated with the path memory in call store. The data in the network map provide the particular line links and trunk links that are available to establish a path through the network as well as those lines already busy. Moreover, a number of busy links may be combined into paths through the network system. Consequently, the network map itself cannot indicate at any time how lines, junctors, and trunks are associated by the existing paths through the network. It is therefore necessary to obtain this information by other areas in call store which collectively are called the *path memory*.

The record of the idle or busy state of all links or paths in an electronic central office is contained in the network map. The idle state is denoted by the binary digit 1, and the busy condition by binary digit 0. Thus, call store and not the switching network must be consulted by central control in order to obtain idle paths in handling a call. A simplified diagram of a network map appears in Fig. 6-7. It shows the possible paths that may be listed in the network map of call store for an intraoffice call from customer A to B. Recall that a binary 1 denotes an idle link and a binary 0 implies a busy link. Therefore, only two idle paths would be available between customers A and B through the line-link network in this example as indicated by the respective arrows on the idle links that are marked 1.

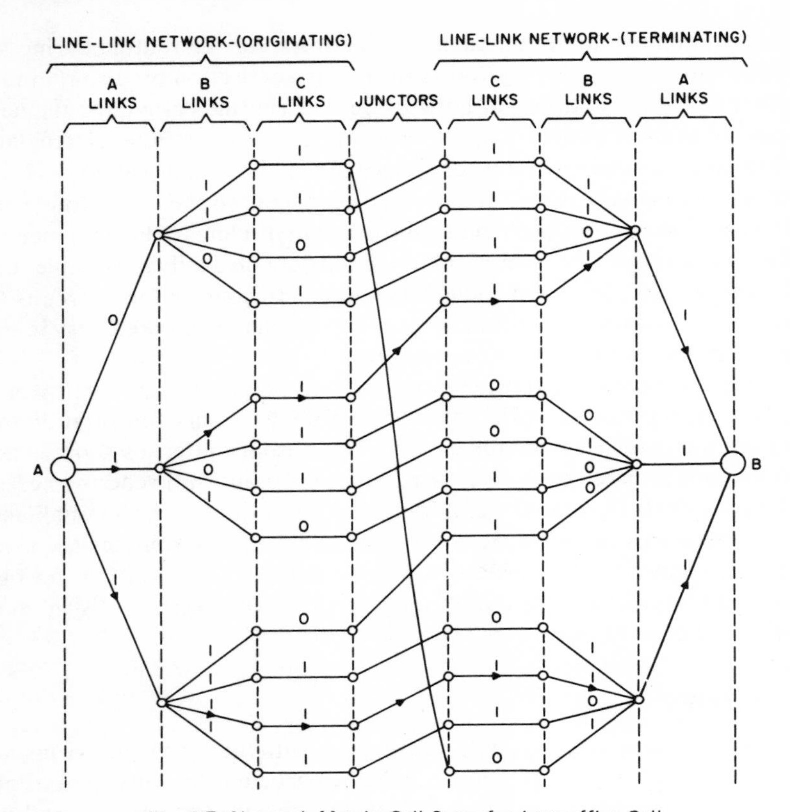


**Fig. 6-7.** Network Map in Call Store for Intraoffice Call

## Functions of Path Memory on Intraoffice Calls

The path memory in call store consists primarily of three types of binary words. One type is permanently associated with each output termination of the trunk-link network. Another is permanently associated with each junctor termination on the line-link network. And several types are assigned as required to the trunk-to-trunk junctors used for intraoffice calls.

Whenever a customer's line is connected through the line-link network to a junctor, the particular path memory's line word associated with the junctor will store the identity of the customer's line. The path memory's trunk word that is associated with the trunk termination in a line-to-trunk connection will contain the identity of the particular junctor involved. In a trunk-to-trunk connection, each of the two path memory trunk words concerned contains the identity of the pair of trunk-to-trunk memory words assigned to the trunk-to-trunk junctor. The trunk-to-trunk memory words in this type of call connection also contain the identity of the two trunks involved and of the trunk-to-trunk junctor.

Figure 6-8 shows the various programs and routines in program store, the related call store hoppers, buffers, registers, and the network equipment with the associated circuits that are usually employed during the ringing and talking connection phases of an intraoffice call. The ringing circuits are scanned every 100 milliseconds, under the control of the ringing trip scanning program, to ascertain whether any called lines have answered. This scanning is accomplished in a manner similar to that described for the customer's line scanning program.

A called-line answer or off-hook indication is first detected by the ringing circuit. This causes the ringing trip scanning program to enter the scanner address data in the ringing trip hopper in call store. When the executive control program schedules the talking connection program to function, each entry in the ringing trip hopper is referred to this program which also may be considered as a ringing answer detection program. The talking connection program then utilizes the path memory idle trunk list to determine the identity of the network appearance of the ringing circuit and of the associated ringing register. Upon receipt of this information, concerning the particular network path involved, the talking connection program proceeds to load this data to an output buffer in call store, designated a peripheral order buffer (POB). The high-priority output program immediately directs this POB to release the ringing register and the related audible ring and ringing circuits.

The talking connection program next makes idle the network paths that were previously used by the audible ring and ringing circuits. It then loads the output buffer or POB with orders to establish the previously reserved talking paths in the line-link network between the originating and called lines. The talking path, which is designated AB and CD in Fig. 6-8, utilizes an associated junctor circuit for operational and control purposes. Note that registers are not associated with intraoffice calls in the talking phase. However, all junctor circuits are scanned every 100 milliseconds under control of the junctor scanning program.

The path memory also provides essential reference data during the ringing stage of an intraoffice call. For instance, assume that the originating customer abandons the call while the called line is being rung. The scanner address of the audible ring circuit would be in the path memory. This information would be translated by a translation subroutine into the identity of the related ringing register and the particular trunk-link terminations of the associated audible ring and ringing circuits. These network terminations, such as those marked P and R in Fig. 6-8, would be subsequently released by timely orders issued by the appropriate output program.

## Program Timetables and Timing Functions

It is seen, from the foregoing, that every action in the process of making a connection between the calling and called lines is controlled by one or

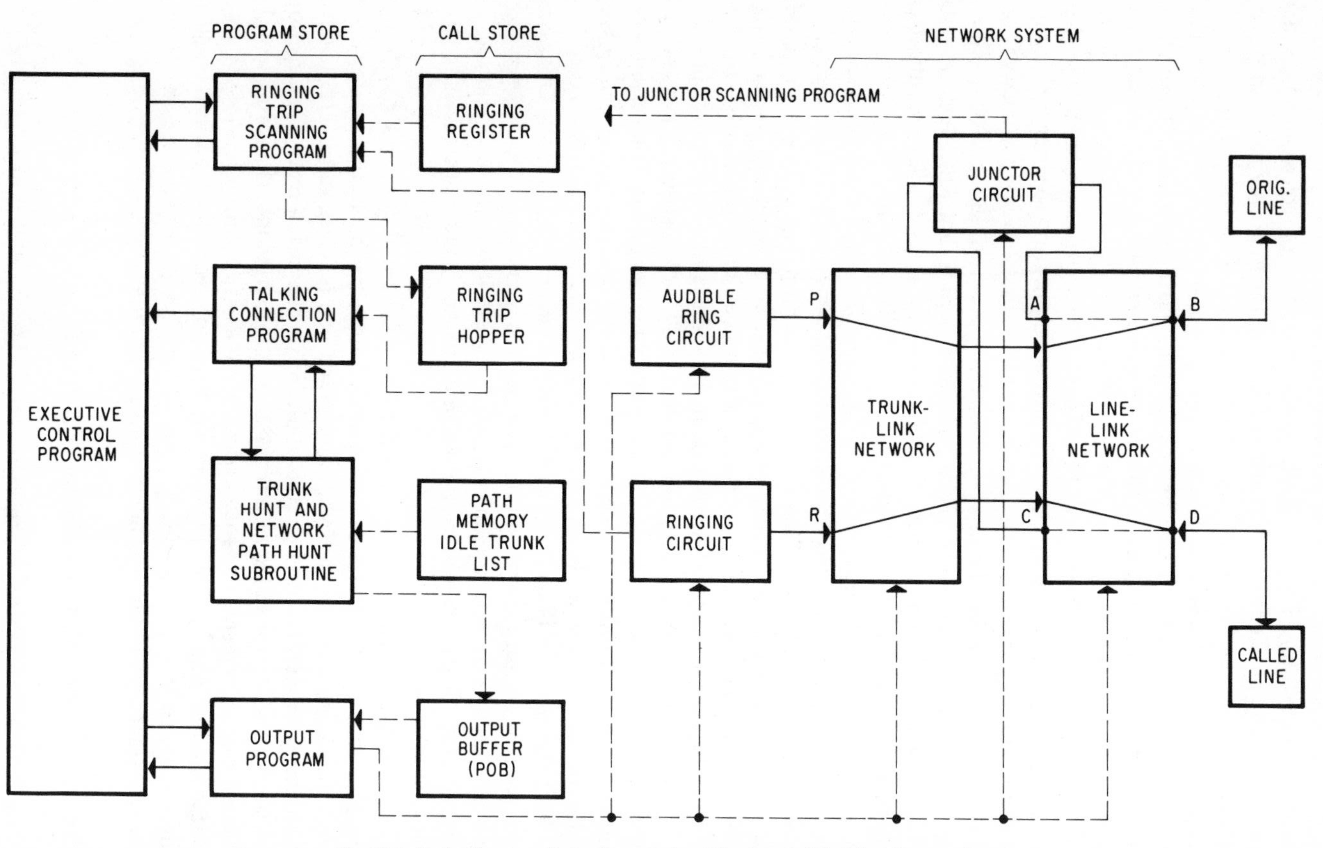


**Fig. 6-8.** Path Memory Functions during Ringing and Talking Connections

more stored programs. The executive control program schedules these functional stored programs in a manner analogous to an orchestra leader who cues in each section of the orchestra at the proper moment. The successful and efficient completion of calls depends on the precise scheduling of the various task programs, especially the input programs.

Note that input signals originate at customers' telephones or in other central offices, while output signals are controlled by specifically assigned programs. Thus, if any input signals are missed or improperly received, calls may go astray or be disconnected. Moreover, many input signals, such as dial pulses, exist only momentarily so that the program assigned to detect them must be scheduled precisely to the millisecond. Input programs, therefore, have precedence over other tasks. Consequently, every 5 milliseconds the particular program in progress is halted by the executive control program to permit an input program, such as line scanning, trunk scanning, or junctor scanning, to take control.

It is necessary that the transition from a call processing routine program to an input detection program occur every 5 milliseconds. It is not feasible, however, for the various task programs to time their own control periods. To do so would necessitate that the programs be written to account for both the random occurrences of inputs and the very accurate timing requirements of the peripheral system equipment. For this reason, the executive control program determines which program will have control at any given time. The programs themselves are not concerned with the timing functions since the 5-millisecond timing pulses are generated by a timing circuit in central control.

The specific timing indicators generally needed by the executive control program are provided by two main timetables in call store. They are designated low priority and high priority with respect to their associated input-output programs. Some examples of the related programs in these timetables are listed in Table 6-2. Each timetable actually is a matrix of binary words and digits. In this connection, the low-priority timetable has a matrix of twenty rows and twenty-three columns. Each row corresponds to a 5-millisecond interval so that the low-priority timetable covers a 100-millisecond period. The high-priority timetable has a matrix of twenty-four rows and twenty-three columns and, consequently, has a 120-millisecond cycle. Each column in both timetables is associated with a particular input-output program, for example, digit scanning, ringing trip scanning, and the ringing circuit answer programs.

The particular program is usually executed when it becomes due. However, it may be rescheduled at a lower priority. To accomplish this, the program would write a binary 1, which is called a *flag bit,* in certain control registers in call store. This flag bit causes the executive control program to schedule the operation of specific task programs at the desired intervals.

**Table 6-2.** Programs with High and Low Priority in Call Store

| Programs | Operating Sequence (milliseconds) |
|---|---|
| **High-priority** | |
| Digit scanning | 10 |
| Abandon and interdigital time-out | 120 |
| Twistor card writing | 5 |
| Call charge monitor tape output | 5 |
| **Low-priority** | |
| Peripheral order buffer execution | 25 |
| Ringing trip scanning | 100 |
| Outgoing trunk scanning | 100 |
| Junctor scanning | 100 |

The high-priority timetable is illustrated in Fig. 6-9. Each row in this timetable matrix corresponds to a binary word and is associated with a particular 5-millisecond interval within the 120-millisecond period of this timetable. The rows in the timetable are read out in turn, one every 5 milliseconds. Whenever a binary 1 appears, the assigned program for the intersecting column is executed. If a binary 0 appears, the corresponding program is omitted

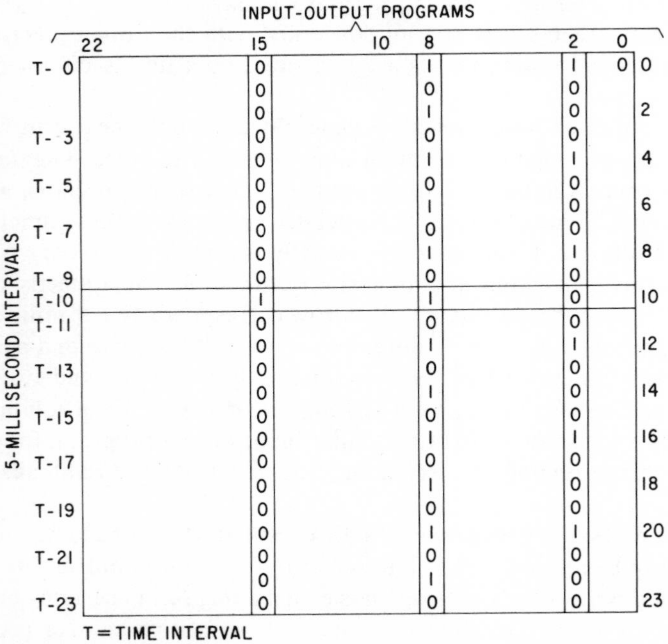

INPUT-OUTPUT PROGRAMS

| 5-MILLISECOND INTERVALS | 22 | 15 | 10 | 8 | 2 | 0 | |
|---|---|---|---|---|---|---|---|
| T- 0 | | 0 | | I | I | 0 | 0 |
| | | 0 | | 0 | 0 | | |
| | | 0 | | I | 0 | | 2 |
| T- 3 | | 0 | | 0 | 0 | | |
| | | 0 | | I | I | | 4 |
| T- 5 | | 0 | | 0 | 0 | | |
| | | 0 | | I | 0 | | 6 |
| T- 7 | | 0 | | 0 | 0 | | |
| | | 0 | | I | I | | 8 |
| T- 9 | | 0 | | 0 | 0 | | |
| T-10 | | I | | I | 0 | | 10 |
| T-11 | | 0 | | 0 | 0 | | |
| | | 0 | | I | I | | 12 |
| T-13 | | 0 | | 0 | 0 | | |
| | | 0 | | I | 0 | | 14 |
| T-15 | | 0 | | 0 | 0 | | |
| | | 0 | | I | I | | 16 |
| T-17 | | 0 | | 0 | 0 | | |
| | | 0 | | I | 0 | | 18 |
| T-19 | | 0 | | 0 | 0 | | |
| | | 0 | | I | I | | 20 |
| T-21 | | 0 | | 0 | 0 | | |
| | | 0 | | I | 0 | | |
| T-23 | | 0 | | 0 | 0 | | 23 |

T = TIME INTERVAL

**Fig. 6-9.** Matrix of High-priority Timetable

during that interval. The readout process is continuous so that when all rows are read out, the cycle starts all over again.

The use of the high-priority timetable, per Fig. 6-9, may be explained by the following examples. Assume that the digit scanning program is assigned to vertical column 8. Since this particular program is executed every 10 milliseconds (see Table 6-2), column 8 will have a binary 1 in every other row as shown in Fig. 6-9.

For another example, assume that the abandon and interdigital time-out program, which is accomplished every 120 milliseconds (per Table 6-2), is assigned to column 15. It will be executed at every row 10 because a binary 1 appears only at the intersection of this specific row and column 15. Moreover, a program to be performed every 20 milliseconds would have a binary 1 in row numbers 0, 4, 8, 12, 16, and 20 as indicated in column 2 of Fig. 6-9.

It is apparent that these timetables cannot be used for sequences that do not repeat themselves either every 100 or every 120 milliseconds. It is necessary, therefore, to utilize a revolving type of counter for other regular repeating time sequences. This type of counter consists of a number of bits in a call store word. The bit at the extreme left side is rotated to the opposite end of the word at regular intervals. Whenever that extreme left bit is a binary 1, the assigned program is performed. The control signal to rotate the extreme left bit is taken from the high-priority timetable.

As an example of this timing sequence, assume that a certain program is executed repeatedly at 45 and 60 milliseconds. This specific sequence may be triggered by a 7-bit revolving counter as represented in Fig. 6-10. Thus, at every third interval of the high-priority timetable (every 15 milliseconds), its output would be applied to this 7-bit counter in order to rotate the leftmost bit. A complete clockwise rotation of the counter from position 0 to position 6 requires 105 milliseconds. Consequently, a binary 1 will appear at the leftmost position (6) every 45 and 60 milliseconds to initiate the execution of the program.

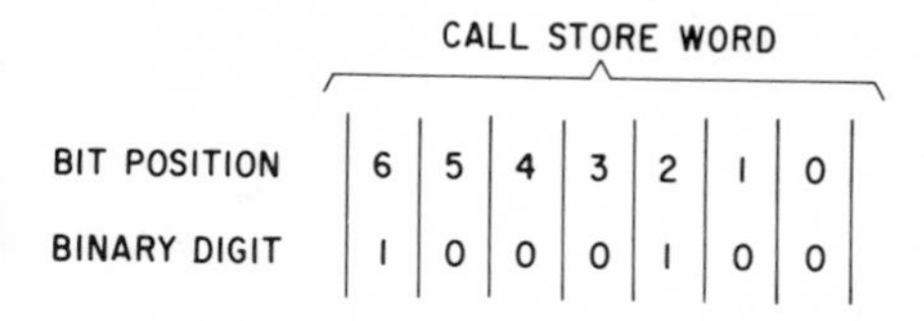


**Fig. 6-10.** Example of Revolving Counter Utilizing Bits in Call Store Word for 105-Millisecond Cycle

## Call Store Register Operations with Stored Programs

The stored program used in the No. 1 ESS and similar electronic switching systems is generic. A single program, therefore, can be utilized to serve many electronic central offices, some having different features or characteristics. As a result, any address, quantity, or option that can vary from office to office must be referred to a specified area of program store, or to call store in the case of recent changes. The job of the task dispenser programs is to examine the data stored in the buffers, and then to decide what particular task programs should be called in to perform the required operations; for example, to direct the scanning of specific customers' lines, or to initiate the ringing connection program.

When this work is finished, the task program involved returns control to the task dispensing program which checks for any waiting interject orders. If none is found, the task dispenser program again examines its associated buffer and the cycle continues. When all task dispensers in one priority class, such as Class B, are completed and their buffers are cleared, the task sequence moves to the next lower priority category, Class C in this case.

All the individual parts of a task program are tied together in the call store memory. Each call in progress is assigned to a call store register which consists of memory areas for the temporary storage of input and control data. Some phases of a call necessitate more memory space than other stages. For instance, seven or ten digits may be stored during digit scanning, depending on whether the call is local or toll. Consequently, there are different-sized registers for the various stages of a call. The program subroutines, however,

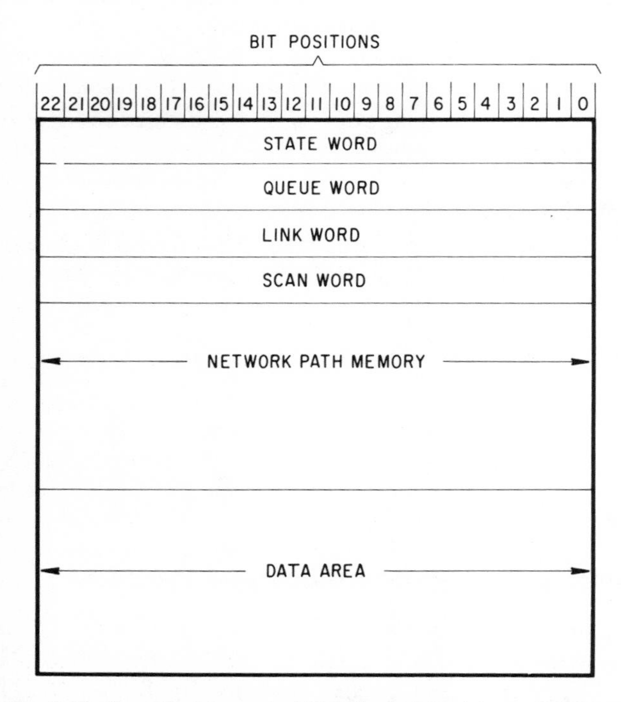


**Fig. 6-11.** Simplified Layout of Call Control Register in Call Store

handle all call control registers in the same manner because these registers utilize a standard format.

Figure 6-11 is a simplified illustration of the layout of a typical call store register. The first four words (each word consists of 23 bits) of any call control register are used for specific storage purposes. The first word is called the *state word* because it identifies the register by its call state and by its function—originating, disconnecting, outpulsing, etc. For instance, when the register receives an input from a task dispenser, the state word selects a group of task programs appropriate to the particular processing stage. The register then selects the exact program by referring to the type of task dispenser program that delivered the input data.

The second word is termed the *queue* because it is used to insert the register on a waiting or timing list. This is often necessary because peripheral equipment, such as an audible ring circuit, or a ringing circuit, may be busy on another call. In this event, when the required ringing circuit becomes available, the queue control program will assign it to the specific register on the queue word. Then, the queue control program will transfer control to the task program that was selected by the state word.

The third word is designated the *link word.* It is used to hold another call control register's address if more than one register is associated with the call. Any number of registers may be linked together by placing the address of the state word of the second register in the link word of the first register, the address of the state word of the third register in the link word of the second register, etc. This chain is closed by putting the state word of the first register in the link word of the last register. Buffer entries to any one register in the chain can be passed to all of them. When a register receives a buffer entry, its state is changed so that future entries will select the task program appropriate to the current state of the call in progress.

The fourth word is the *scan word* which is not always used. Its purpose is to link a call store register to a scanning register when the call control program has requested the scanning of some particular points in the system. The next group of words in the register is set aside for the network path memory. It may differ in length of words from one type of call control register to another. The storage area following the path memory is the *data area.* Its size depends upon the amount of data required by the call control programs.

It is obvious that the stored program is the operating intelligence of the electronic switching system. The program structure is determined to a large extent by the call handling requirements of the electronic switching system, and the fact that it must respond to service requests as they occur, that is, in real time. Specific data such as equipment quantities and translations, which vary from office to office, are not written into the generic program. However, they can be easily added to the memory system in any office. The program can be readily modified to incorporate new service features or changes. Consequently, it may be claimed that the No. 1 ESS and related electronic switching systems can be arranged to control telephone services and features that have not yet been foreseen.

## Questions

1. What major stored program elements are contained in program store? Approximately how many instructions comprise the stored program and what percentage is devoted to call processing and related functions?
2. Name the four functional groups of the stored program and briefly describe their functions. Which one is the main program?
3. What three elements in call store are associated with stored program control operations? Briefly describe their functions.
4. What are the two major program interrupt categories? Which has the higher priority? What is the number of interrupt levels and how are they designated? Which one is the lowest level and what function does it perform?
5. What are task dispenser programs and how are they scheduled?
6. What functions are performed by a translation program? What is the designation of a translation program that always returns control to its task program?
7. Where is the network map located and what does it contain? What indicates the state of a path in this map?
8. Describe the main elements of the path memory in call store.
9. What major timing indicators are provided for the executive control program? What are their designations? Briefly describe their structure and timing periods.
10. What are the designations of the first four binary words in a call control register in call store? Briefly describe their functions.

# 7 SWITCHING NETWORKS, MASTER CONTROL CENTER, AND POWER SYSTEM

## Switching Network Precepts

The step-by-step, panel, rotary, and crossbar electromechanical systems, briefly described in Chapter 1, utilize various mechanical and electromechanical devices to accomplish their switching functions. For instance, in step-by-step central offices, the selector and connector switches themselves make the required interconnections between the calling and called lines in sequence with the dialed digits. In panel offices, the switching operations are performed by selector brushes which glide over terminals on vertical banks under control of a register-sender. In crossbar systems, the marker first checks the S (sleeve) leads to find a set of idle switching links and an idle trunk. It then closes the corresponding crosspoints on selected crossbar switches to accomplish the network connections.

In the original design for an electronic switching system, the use of electronic means for the actual switching functions was also considered. In fact, solid-state devices and gas vacuum tubes were employed as the switching network in some of the first laboratory and trial installations. There were, however, some serious drawbacks. For instance, gas tubes were unable to handle the high amplitude 20 Hz ringing signals.

Solid-state devices such as diodes and transistors likewise have deficiencies when used for switching telephone lines and trunks. These electronic devices can readily pass current in the forward direction and present a high resistance to reverse current flow. However, they can also rectify voice-frequency signals and, therefore, may impair transmission by introducing noise in other parallel paths. This condition is illustrated in Fig. 7-1. It shows six telephone line circuits, any one of which may be connected or switched to the outgoing trunk circuit, To, Ro. This switching may be accomplished by applying a suitable bias current to the respective pair of switching diodes associated with the particular line circuit.

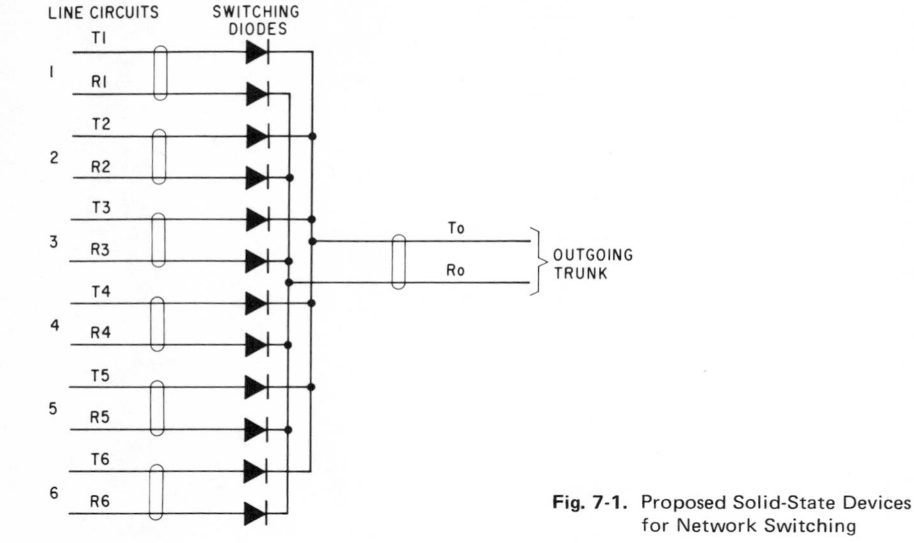


**Fig. 7-1.** Proposed Solid-State Devices for Network Switching

For example, by applying the correct bias to the switching diodes of line circuit 4, only this line will be connected to the outgoing trunk. The diode outputs of the other lines, however, are in parallel with each other as well as bridged to line 4. Thus, noise voltages may be introduced into line 4 and its connected outgoing trunk circuit, thereby downgrading voice transmissions. This condition would be aggravated as the number of parallel paths is increased.

Moreover, field trials and laboratory tests have indicated the need of the physical opening of unused parallel paths in order to ensure good transmission, and to provide for compatibility with existing telephone plant and equipment. Consequently, mechanical devices, such as relay contacts or crosspoints on crossbar switches, were used rather than electronic means. A high-speed reed switch with balanced metallic contacts was subsequently developed by the Bell Telephone Laboratories to perform the switching network interconnections in electronic offices. It was designated the *ferreed switch.*

## Ferreed Switch

The ferreed switch is the basic interconnection device first utilized in the switching network of the No. 1 ESS and similar electronic central offices. It is a two-wire magnetically latched device consisting of two miniature magnetic reed switches, each sealed in a glass tube. The two sealed reed switches are mounted between two rectangular remendur plates. One reed switch controls the tip conductor and the other controls the ring conductor of the two-wire transmission path through the switching network.

Remendur is a two-state magnetic alloy which can be changed very quickly from one magnetic state to the other by means of very short current pulses through a surrounding wire coil. This particular magnetic material also has a high remanence, that is, it will stay magnetized until another current pulse switches it back to its previous state. This switching action can be accomplished in a millisecond or less. The ferreed, therefore, is many times faster than other electromechanical switching devices.

Two types of ferreed switches are provided for the switching network. They are termed the *crosspoint ferreed* and the *bipolar ferreed.* The crosspoint ferreed, or ferreed as it is usually called, supplies the crosspoint contacts for the T and R conductors in the eight stages of switching provided in paths through the line-link and trunk-link networks. The bipolar ferreed is utilized for two main functions. It disconnects a customer's line circuit from its associated scanner ferrod (see Fig. 2-9). This function is similar to that of the cut-off (CO) relay in the step-by-step and panel central offices. The bipolar ferreed switch is also associated with junctor switching frames. It provides access to circuits used to test a line or trunk for false cross and ground (FCG) conditions, and for no-test and restore-verify requirements.

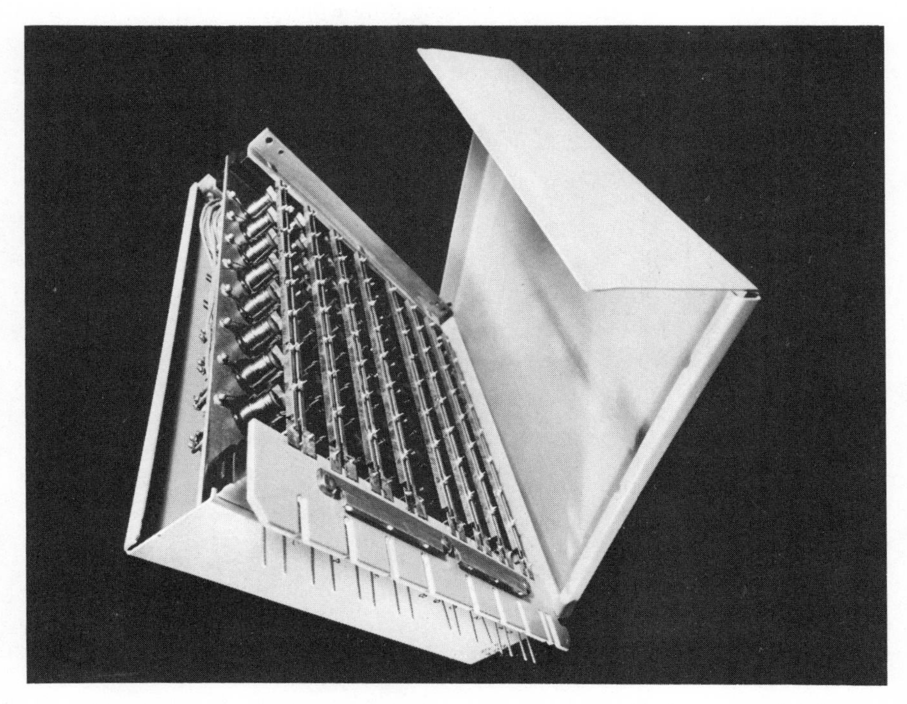

**Fig. 7-2.** Eight-by-Eight Crosspoint Ferreed Switch Assembly (Courtesy of Bell Telephone Laboratories)

The crosspoint ferreeds are usually assembled in a square array of sixty-four switches consisting of eight columns and eight rows. Figure 7-2 is a photograph of the basic 8 by 8 two-wire crosspoint array of ferreed switches. The reed contacts are connected to horizontal and vertical straps which form the network multiple. When connected by the reed contacts, these horizontal and vertical multiples provide the T and R paths through the ferreed switch.

Eight bipolar ferreeds usually are assembled to form a switch assembly. Both ends of all crosspoints may be brought out to terminals in order to provide eight individual crosspoints in one assembly. One side of all crosspoints also may be strapped together to provide a 1 by 8 type of switch.

## Operation of Ferreed Crosspoint Switch

The main components and the construction of the crosspoint ferreed are illustrated in Fig. 7-3. The two remendur plates are divided magnetically into two halves by the horizontal steel shunt plate which is positioned at their centers. The two control windings are wound on a form between the two remendur plates. Each control winding has about sixty turns, with forty turns on one-half of the ferreed assembly opposing twenty turns on the other half of the switch. This arrangement is represented by windings 1 and 2 in Fig. 7-3.

When only one winding is energized by a current pulse, the two halves of the remendur plates are magnetized in *series-opposition* (the adjacent ends

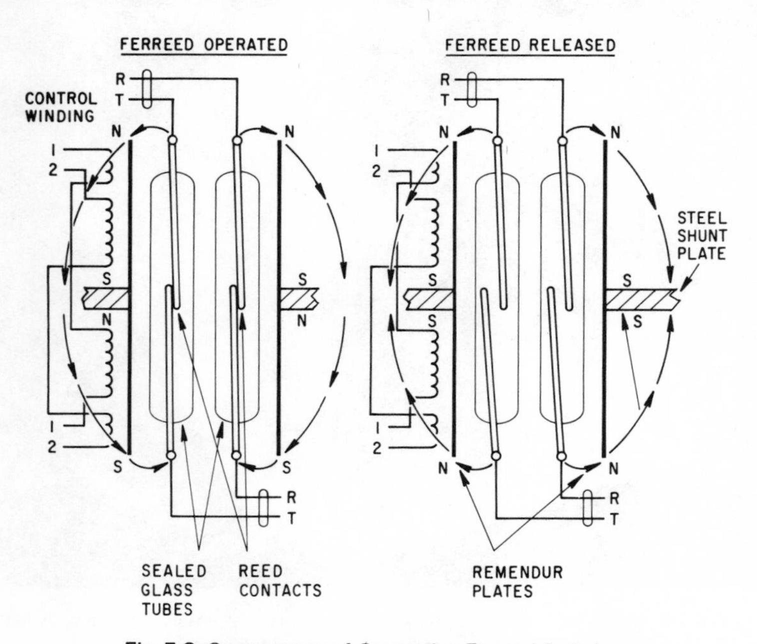


**Fig. 7-3.** Components of Crosspoint Ferreed Switch

have the same magnetic polarity). As a result, there will be very little magnetic flux through the reeds and their normal tension will open the contacts. This condition is portrayed in the right-hand sketch in Fig. 7-3. Now if both control windings are pulsed simultaneously with equal currents, both halves of the remendur plates will be magnetized (*series-aiding*) and the reed contacts will close. The left-hand sketch in Fig. 7-3 illustrates this state. The remendur plates retain their magnetic polarity after the current pulses cease so that the reed contacts remain closed. Note that ferreeds do not require power to keep their contacts closed.

The reed contacts of all ferreed switches always open and close under "dry" circuit conditions. That is, there is no battery potential applied to the line, path, or trunk during the switching interval. Thus, the reed contacts do not open or close current-carrying conductors. Relays in the associated line scanner, trunk, junctor, and service circuits function after the switching interval to apply battery potential to the corresponding T and R conductors. This method ensures a very long life for the miniature ferreed switches.

## Operation of Bipolar Ferreed Switch

The bipolar ferreed is constructed similarly to the crosspoint ferreed except that the two sealed reed switches are mounted between two magnetic alloy rods. One rod is a permanent magnet and the other is a semipermanent

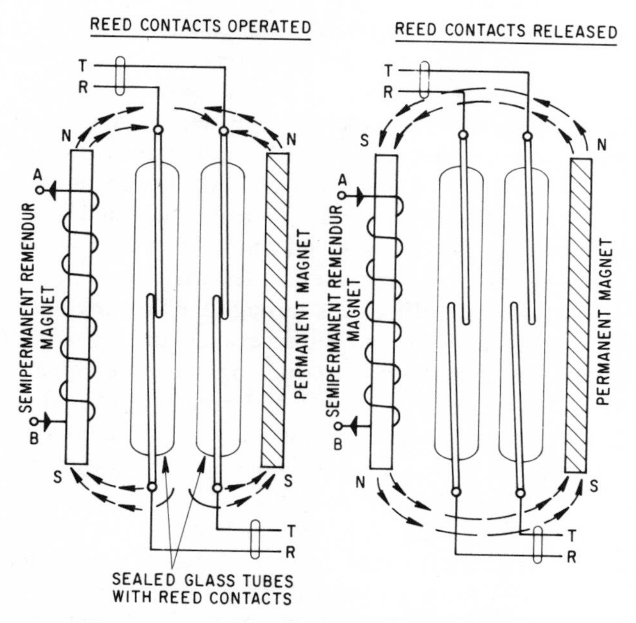


Fig. 7-4. Components of Bipolar Ferreed Switch

magnet made of remendur. The magnetic polarity of the remendur rod is controlled by the direction of the current pulse sent through its winding.

The sketches in Fig. 7-4 illustrate the construction of the bipolar ferreed switch. When a current pulse is sent through the control winding in the BA direction (left-hand sketch) the generated magnetic flux will add to the flux from the permanent magnet rod, and both will pass through the reeds. The reed contacts, consequently, will close because of the greatly increased magnetic flux.

When the control winding is pulsed in the opposite or AB direction (right-hand sketch), the resultant magnetic polarity of the semipermanent magnet will attract the flux from the permanent magnet. As a result, almost all magnetic flux will bypass the reeds in the sealed glass tubes. The normal tension of the reeds, therefore, will cause their contacts to open. In either case, when the current pulses ceases, the semipermanent remendur rod retains the resulting magnetic state. Thus, the bipolar ferreed remains operated or released after the current pulse.

## Switching Network Elements

The switching network is of the octal type consisting of eight stages of ferreed switches. The crosspoints of these ferreeds establish two-wire metallic paths, commonly designated the tip (T) and ring (R) conductors, for the transmission of voice and control signals. This switching network, besides interconnecting customer lines and trunks, also connects lines to junctor circuits and to various service circuits (dial pulse receivers, ringing and tone circuits, and signal transmitters) during the processing of a call.

The main equipment units of the switching network consist of a number of line-link and trunk-link networks which are interconnected through the junctor grouping frame. The number of such networks depends upon traffic requirements of the particular electronic central office, and may vary from one to sixteen frames for the line-link and trunk-link networks, respectively. The junctor grouping frame also provides the terminations for the junctor circuits which are used for interoffice calls through the line-link network.

The major elements of the switching network and the path of an interoffice call through the eight stages of switching are represented in Fig. 7-5. The path is made up of links connected by ferreed switches that were previously described. Each link is essentially a two-wire path consisting of the tip and ring conductors. It is divided into two stages for switching purposes, which are designated 0 and 1 as shown in Fig. 7-5.

In the line-link network, the line switching frames perform the first two stages of switching by connecting the customer lines on their A link inputs to the B links associated with the outputs of the A links. The junctor switching frames, in both the line-link and trunk-link networks, likewise perform two stages of switching. Their C links interconnect the B links with the junctor

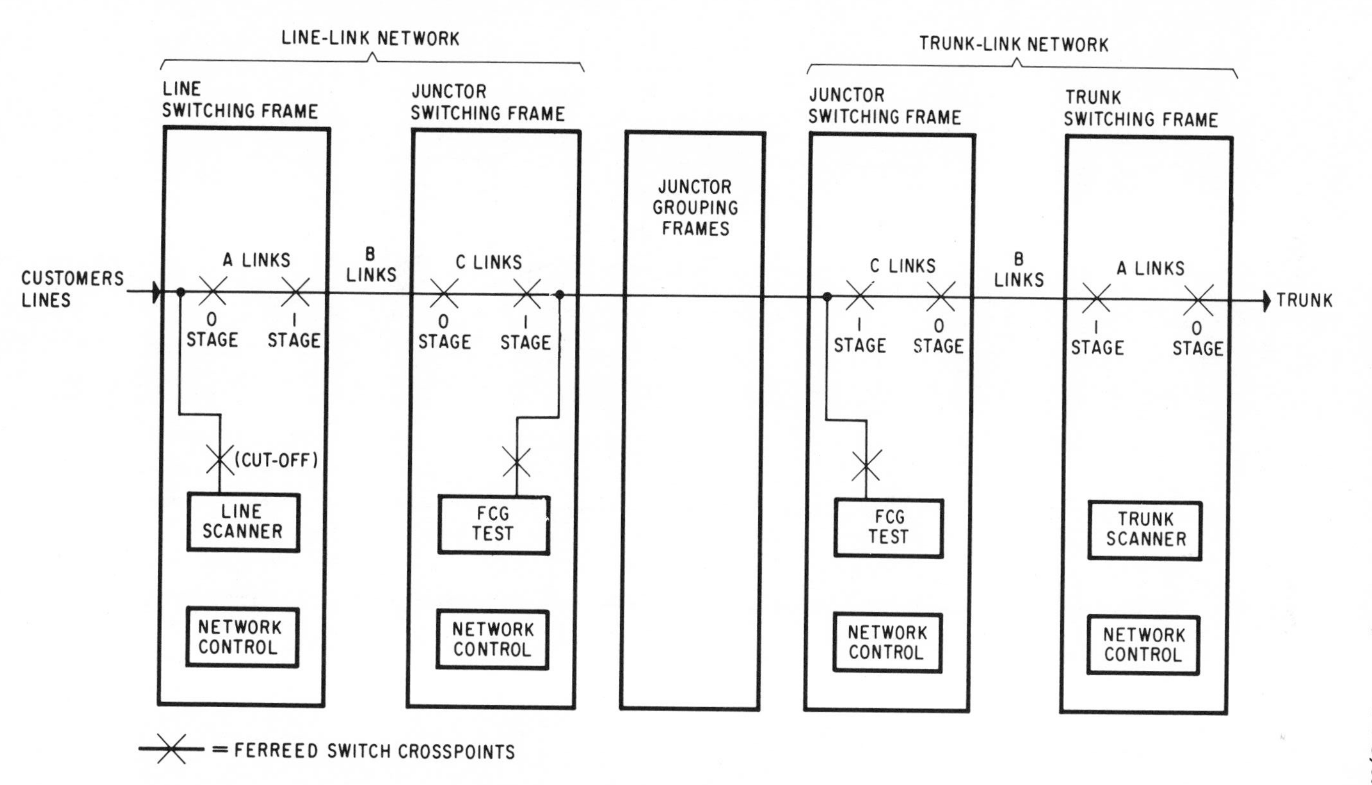


**Fig. 7-5.** Major Elements of Switching Network

grouping frame on interoffice calls in order to provide a path to the trunk switching frame. Two stages of switching also are completed by the A links in the trunk switching frame to connect to the selected outgoing trunk. The FCG test refers to the false cross and ground tests previously explained. These tests are made on a segment of a path before it is closed through the network, to ensure that the conductors are not crossed or grounded.

The line switch frames also accomplish the traffic concentration function of the network. Two different concentrator ratios are usually provided by the line switching frames between the inputs from customer lines and the out-

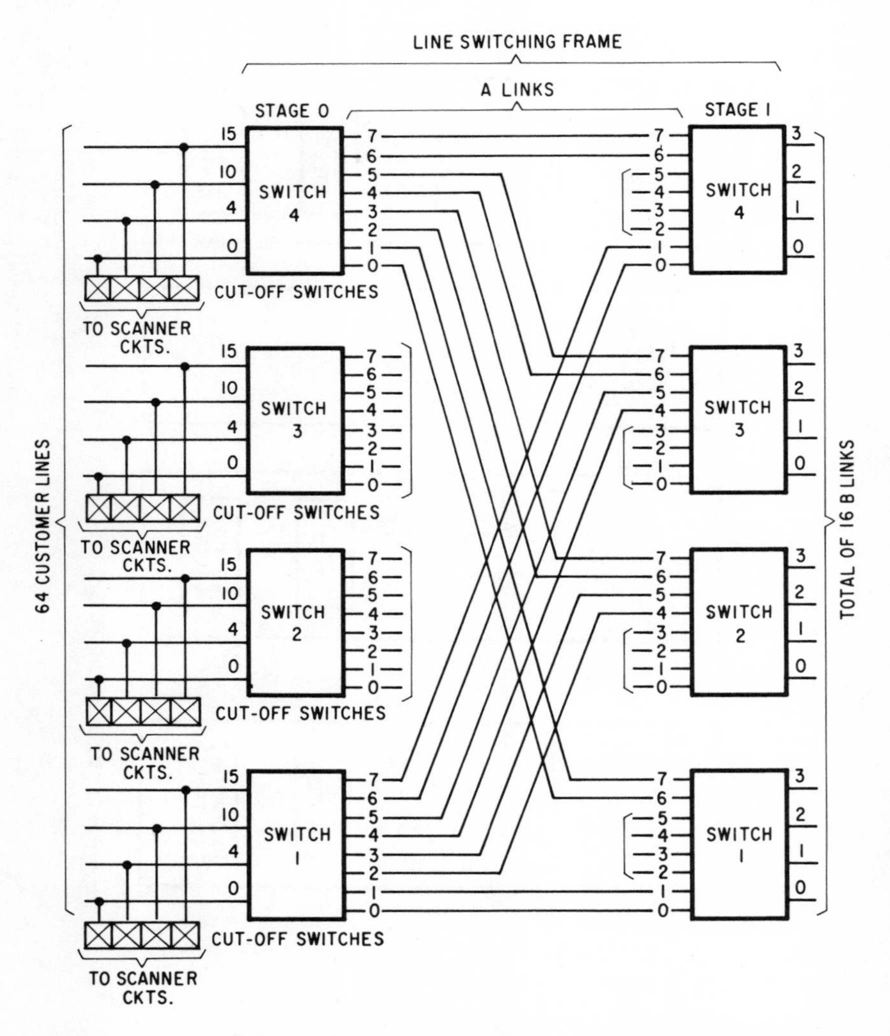


**Fig. 7-6.** Simplified Block Diagram of 4 to 1 Concentrator Line Switching Frame

puts to the B links. These ratios are 4 to 1 and 2 to 1. The 4 to 1 concentrator line switching frame has sixteen line concentrators. Each concentrator interconnects sixty-four customer lines to sixteen B links through the two switching stages designated 0 and 1 as shown in Fig. 7-5. Thus, one complete line switching frame terminates 1,024 customer lines on its input and 256 B links on its output.

The 2 to 1 line switching frame also has sixteen line concentrators. Each concentrator interconnects thirty-two customer lines to sixteen B links through two switching stages in a similar fashion as in the case of the 4 to 1 concentrator. Therefore, 512 customer lines may be terminated on the inputs of this type of line switching frame for connection to 256 B links. Other concentrator ratios, such as 3 to 1, 5 to 1, 6 to 1, 7 to 1, or 8 to 1 may be obtained by using different combinations of junctor switching frames with the line switching frames.

Figure 7-6 is a simplified block diagram of a line switching frame that is arranged for a 4 to 1 concentrator ratio. Note that a group of sixty-four customer lines is assigned to four input or stage 0 switches, utilizing sixteen lines per a switch. Each switch is an array of sixteen crosspoint ferreeds. The wiring between stages 0 and 1 is designated the A links. Each of the customer lines on a stage 0 switch can be connected to only four of the eight outputs of the switch. These four outputs of the switch (A links) are wired, one each, to the four switches in stage 1. Figure 7-6 also illustrates examples of wiring between some of these switches. Each line in the drawing represents a tip and ring pair. Access to all B link outputs, consequently, is obtained with a minimum number of crosspoint ferreeds per A line.

Note that each customer's line also connects to a pair of bipolar ferreeds or cut-off switches. These cut-off switches provide the connections to the line scanner circuit which initiates requests for service. Control equipment in the line switching frame governs the operation of these bipolar ferreed switches.

## Trunk and Junctor Switching Frames

The trunk and junctor switching frames are similar in construction except that the trunk switching frame is not equipped with bipolar ferreed switches for access to the FCG test circuit. Both types of switching frames perform two stages of switching. The trunk switching frame interconnects trunks and service circuits on its inputs (A links) with the B links on its outputs. The associated junctor switching frame interconnects the B links with the two stages of its C links. These three interconnecting links or paths are completely within the trunk link network as shown in Fig. 7-5 and 7-6.

Both trunk and junctor switching frames are usually arranged for a 1 to 1 concentrator ratio utilizing four octal grids. Figure 7-7 is a simplified block diagram of a typical junctor switching frame. This diagram is also applicable

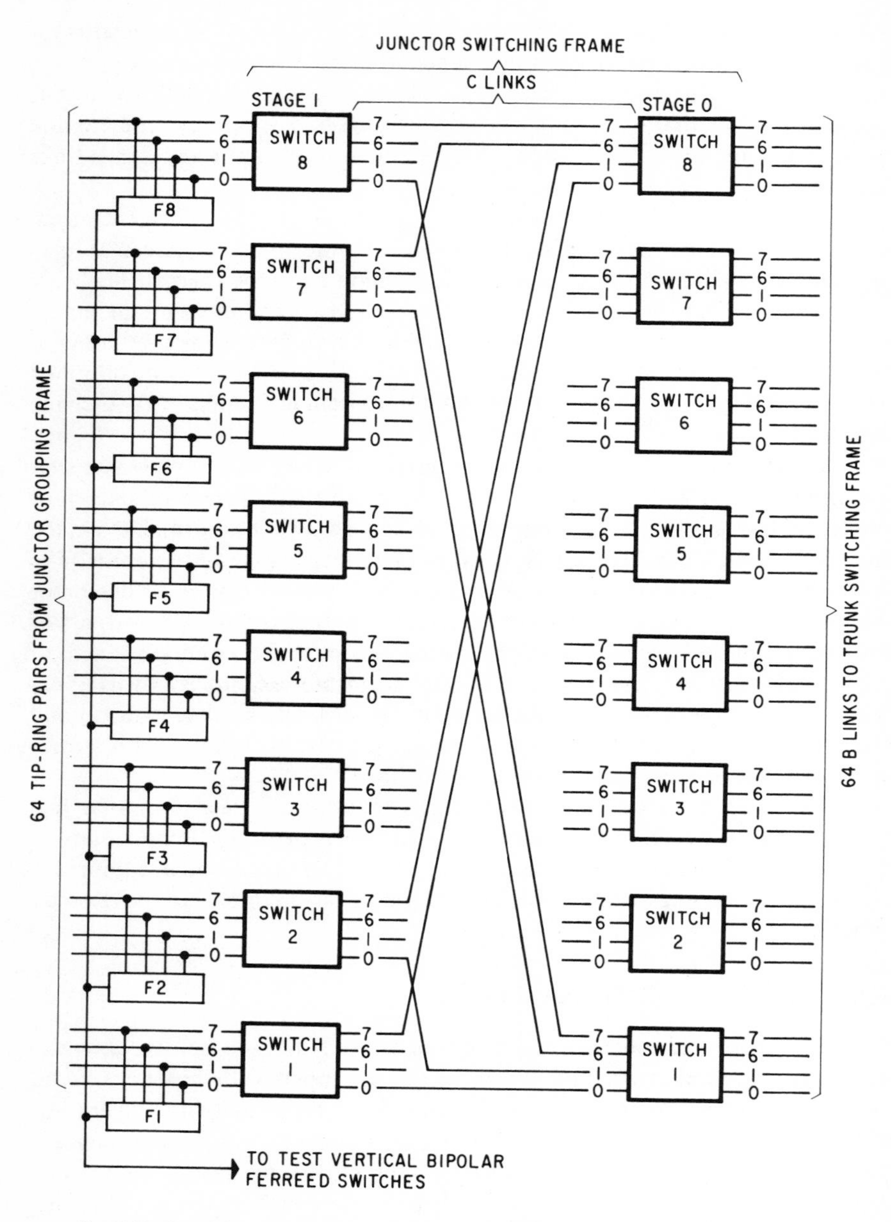


**Fig. 7-7.** Simplified Block Diagram of Junctor Switching Frame in Trunk Line Network

to its associated trunk switching frame. The sixty-four inputs or B links are assigned to eight switches in stage 0, and the sixty-four outputs are connected to eight switches in stage 1 by the C links. Each switch handles eight links or tip and ring conductor pairs. The interstage wiring connects each input switch to each of the eight output switches. Therefore, each input or B link will have access to every one of the sixty-four output terminals through a single path defined by a particular C link.

## Junctor Grouping Frame

The line-link and trunk-link networks are interconnected through two or more junctor grouping frames. The junctor grouping frame actually provides an interconnecting path between the junctor switching frame of the line-link network and the junctor switching frame of the trunk-link network on all interoffice calls. This path is shown in Fig. 7-5. For intraoffice calls, the junctor grouping frame connects a junctor circuit associated with the junctor frame to the C-link output of the line-link network. Likewise, the junctor grouping frame provides trunk-to-trunk interconnections on calls routed from distant central offices through the particular electronic switching office. The block diagram in Fig. 7-8 illustrates the various interconnections provided by the junctor grouping frame.

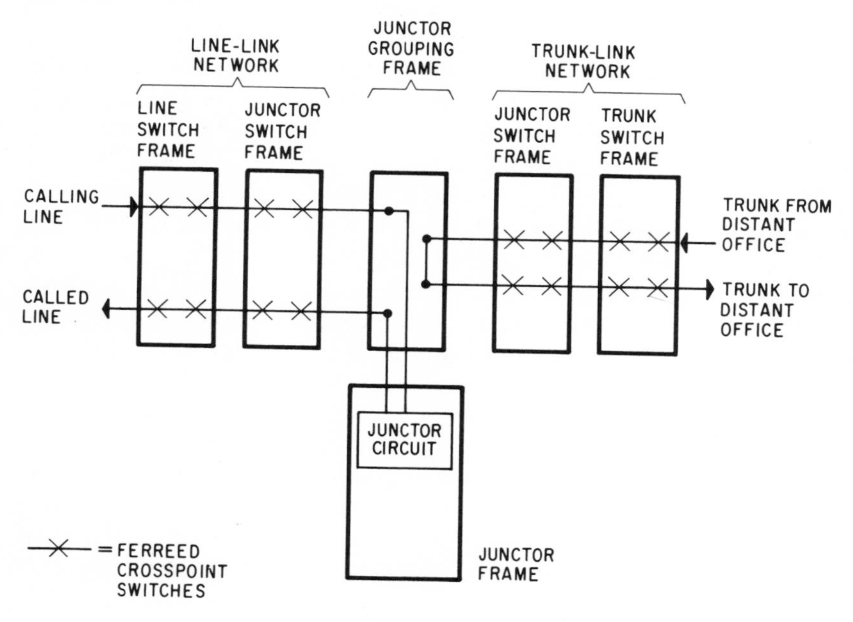


**Fig. 7-8.** Interconnections Provided by Junctor Grouping Frame for Intraoffice Calls and Trunk-to-Trunk Connections

The junctors from the line-link and trunk-link networks are arranged in subgroups of sixteen tip and ring conductor pairs. These sixteen pairs are connected via terminal strips on the rear of the junctor grouping frame, either to a connector or to a plug-ended patch cord on the front of the frame. This arrangement is illustrated by the photograph in Fig. 2-7. Thus, any desired interconnection pattern may be made between the line-link and trunk-link networks by the jack and plug system installed in the junctor grouping frame. Equivalent interconnection patterns in electromechanical central offices would necessitate substantial rewiring work.

## Establishing Network Paths

Paths through the line-link and trunk-link networks are created by a combination of links and stages of ferreed switches operated under the direction of central control. These links and associated wiring form a transmission or talking path when the appropriate crosspoint contacts on the related ferreed switches in the network are closed.

Each switching frame in the network contains two controllers which are operated by the central pulse distributor (CPD) under the direction of central control. Every network controller governs the operation of one-half of the ferreed switches in the particular switching frame. In normal operation, each network controller functions with its assigned group of ferreed switches. Thus, two simultaneous path selections can be made within a switching frame, one under the direction of each controller. If a network controller should fail, its mate will take control of both groups of ferreed switches. In this event, only one path selection will occur at a time within that particular switching frame.

To explain the operations involved, let us first consider how paths in the network are produced for an intraoffice call from customer line A to line B as shown in Fig. 7-9. When line A is idle (on-hook) prior to initiating the call, its assigned contacts on the crosspoint ferreed in stage 0 of the line switching frame are open. At the same time, the cutoff contacts on the bipolar ferreed associated with line A's ferrod sensor will be closed. (Refer to Fig. 2-9 for a schematic diagram of a line's ferreed contacts and its ferrod sensor.) Concurrently, the line scanner circuit checks periodically for an off-hook indication to start the necessary actions when customer A originates a call.

As soon as the call to line B has been dialed by calling customer A, central control will initiate the setting up of a path through the line-link network to interconnect the calling and called lines. For this purpose, central control maintains a current record of the busy-idle states of all network links on a network map, as per the example shown in Fig. 6-7. Also, central control keeps a record of the end termination of all network paths in use or reserved within the path memory of call store as explained in Chapter 6 and Fig. 6-8.

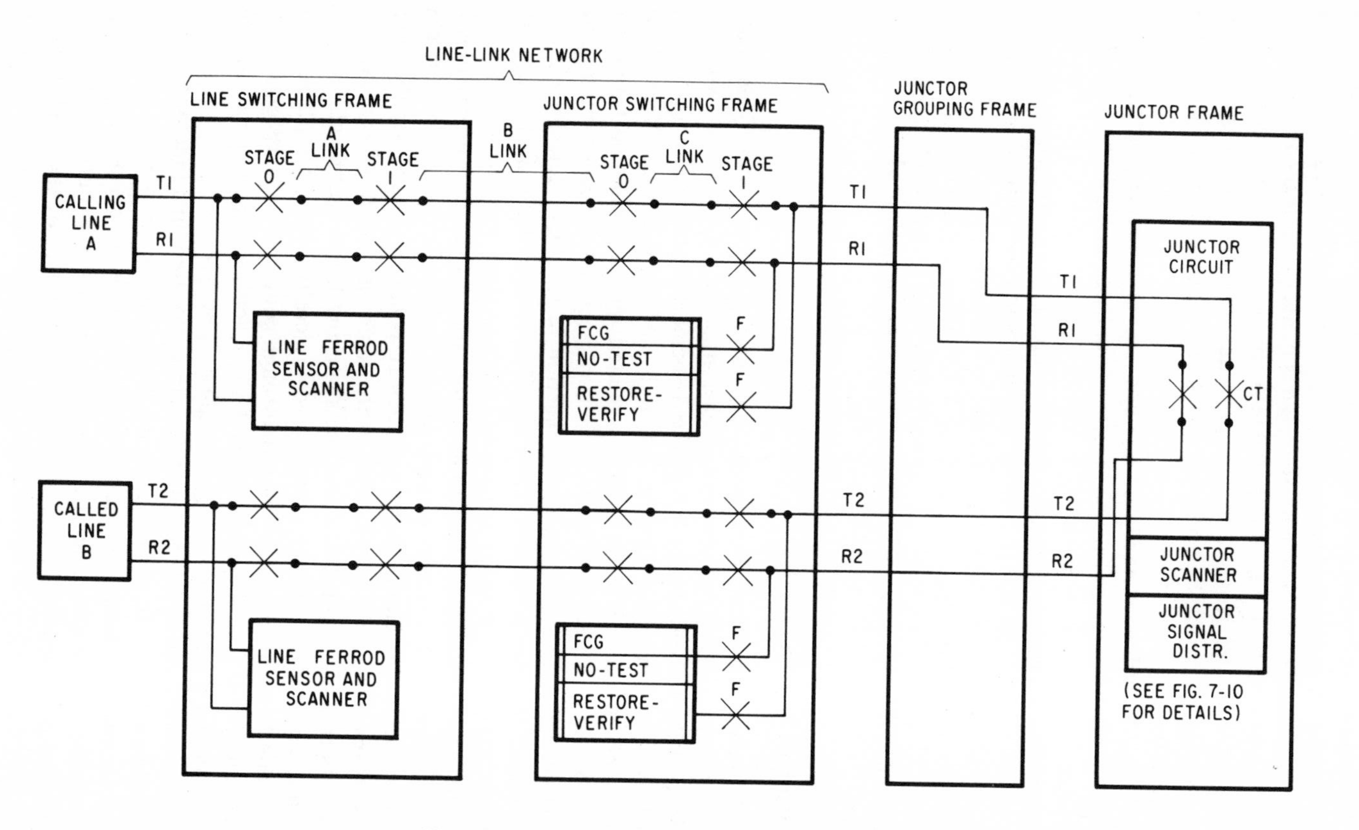


**Fig. 7-9.** Network Paths Established for an Intraoffice Call

The network map in call store is examined whenever a path is to be selected between a line and junctor switching frame, or between a trunk and junctor switching frame. The path selection is made by central control on the basis of the busy-idle states of all possible A, B, and C links between the line and a selected junctor group. The same procedure is also followed in choosing a path through the trunk-link network to a selected trunk in the case of an interoffice call.

After the path is selected, the network map in call store is updated and a record of the line junctor (or line junctor trunk association in the case of an interoffice call) is entered in the path memory of call store. Central control next prepares appropriate network instructions. These instructions or work lists, which include the identity of the line link and junctor frames involved, are recorded in a particular peripheral order buffer (POB) in call store. One POB, as explained in Chapter 6 and Fig. 6-6, is associated with each call requiring a connection through the network. The aforementioned instructions subsequently are read out of the POB at 25-millisecond intervals for transmission to the required network controllers.

## Transmission Paths for Intraoffice Calls

A partial path first is set up through the line-link network. This is accomplished by closing the ferreed switches in stage 1 of the line switching frame, and the ferreeds in stage 0 of the associated junctor switching frame. This is indicated in Fig. 7-9 by the T-1, R-1 and T-2, R-2 leads. Next, the F contacts of the bipolar ferreeds in the related junctor switching frame are closed to connect these same leads to their respective false cross and ground (FCG) verticals. This permits the connection of the FCG detector circuits which check the conductors (T-1, R-1 and T-2, R-2) in the paths for any foreign potential, false crosses, or grounds.

Assuming that the FCG path tests are satisfactory, central control will direct that the cutoff contacts associated with the line ferrods of both lines A and B in the line switching frame, and their respective F contacts in the junctor switching frame, be opened. Also, at the same instant, the crosspoint contacts of both lines in stage 0 of the line switching frame will be closed. The last step, directed by central control to complete the talking path between lines A and B, is the closure of the cut-through contacts in the assigned junctor circuit. Note that these operational steps apply simultaneously for the paths assigned to both the calling and called lines.

The schematic of a typical junctor circuit located on a junctor frame is illustrated in Fig. 7-10. This junctor circuit furnishes the battery power required for voice transmission and signaling. Separate battery potential is supplied to the calling and called telephone stations for this purpose. The cut-through latching relays (CT-1) and (CT-2) are operated by the junctor signal distributor circuit after all ferreed switches in the line-link network are closed

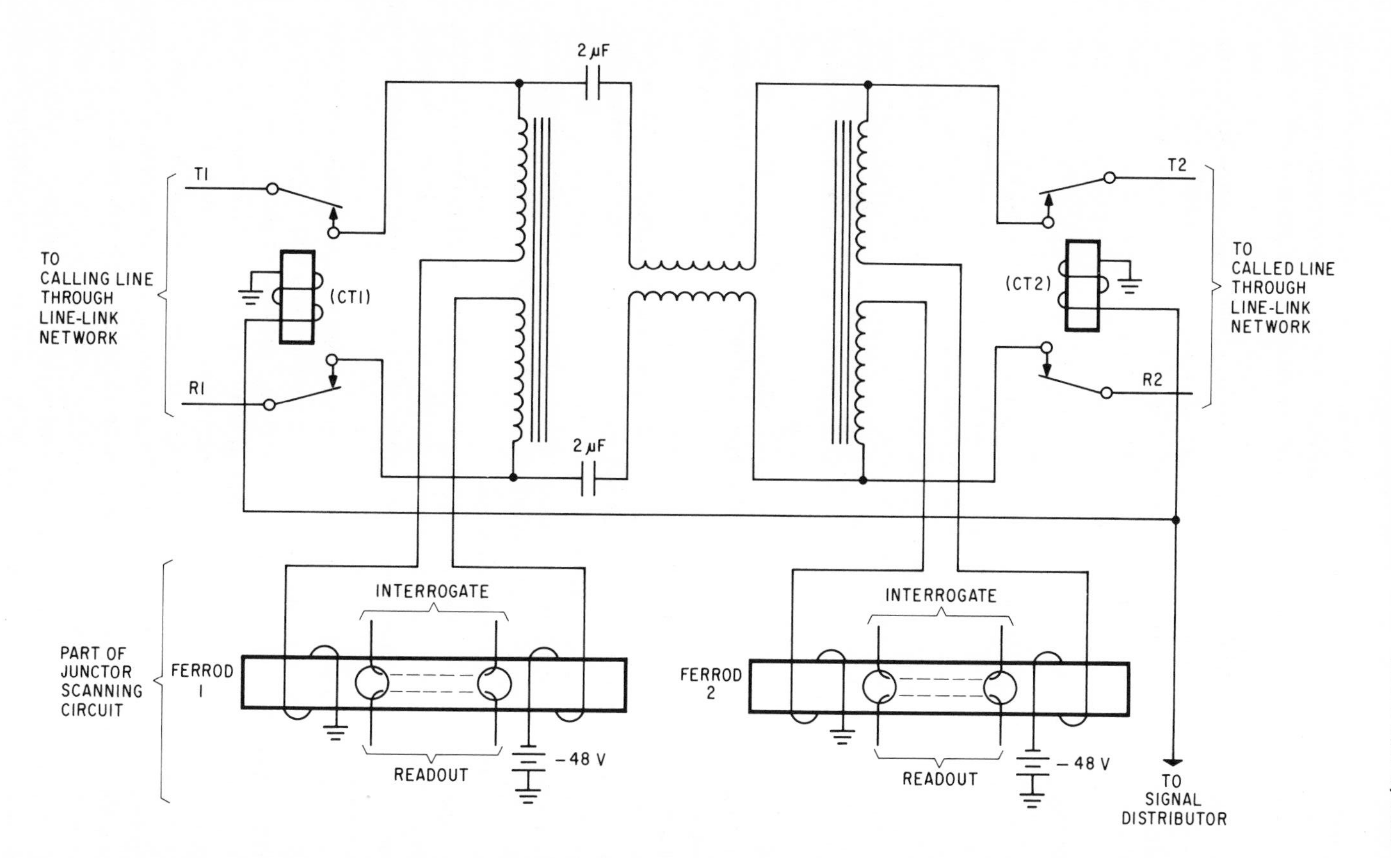


**Fig. 7-10.** Junctor Circuit Used to Supervise Calling and Called Customer Lines on an Intraoffice Call

to complete the path between lines A and B. In this manner, no battery potential is present on the ferreed crosspoints until after the contacts are closed. On disconnection, these latching relays are released before the crosspoint contacts are opened.

The 2 $\mu$F capacitors in the transmission path serve to bypass the speech currents. The inductors in series with the T-1, R-1 and T-2, R-2 conductors act as audio-frequency choke coils to keep out the speech currents while allowing the passage of the 48 volt direct current.

## Call Supervision by Junctor Circuit

Ferrods 1 and 2 supervise the calling and called line circuits, respectively, for a change of state. The presence or the flow of direct current through the control windings of the ferrods will cause a binary 0 to be read out by the junctor scanning circuit whenever the interrogate winding is momentarily pulsed. The absence of current flow results in a readout of a binary 1.

These binary readouts advise central control of the status of the call. For instance, readouts of binary zeros would indicate that both parties are connected with each other and conversing. Readouts of binary ones would signal that both parties have hung up. Readouts of a binary 0 from ferrod 1 and a binary 1 from ferrod 2 would show that only called line B has disconnected. Similarly, readouts of binary 1 from ferrod 1 and a binary 0 from ferrod 2 denote that only the calling line A is on-hook.

When calling line A hangs up at the end of a conversation, a binary 1 would read out from ferrod 1 by the junctor scanning circuit to central control. The release of the connection is initiated by the release of the cut-through latching relays (CT-1) and (CT-2) in the junctor circuit as depicted in Fig. 7-10. The line-link network paths for the calling and called lines, however, remain closed at this time. Next, central control directs the closure of the bipolar ferreed's cutoff contacts for both lines A and B in their respective line ferrod sensor and scanner circuits in the line switching frame. Simultaneously, central control causes the closure of the corresponding F contacts in the junctor switching frame.

The restore-verify circuit will be connected to each of lines A and B by the closure of the F contacts. This circuit connects a resistor across the respective T-1, R-1 and T-2, R-2 leads of the network paths to simulate the origination of a call. This condition will be observed by the line ferrod sensors and scanners of lines A and B, and central control will be informed of the resultant satisfactory path tests. As a result, central control will direct that the crosspoint contacts in stage 0 of the line switching frames be opened to complete the disconnection of the call and to make the network paths available for other calls.

It is possible to monitor conversations at any time during a call by closing the F contacts in the junctor switching frame and connecting the no-test circuit shown in Fig. 7-9. The no-test circuit, however, can be activated only from the master control frame. It is utilized for testing purposes and to check for busy-line conditions.

## Improved Remreed Switching Network

An improved magnetic reed switching device has been developed by the Bell Telephone Laboratories to replace the ferreed switch in electronic switching offices. This new switching device is called a *remreed*. It utilizes all electronic controls, has greater reliability, and achieves a 4 to 1 space reduction over the present ferreed switching networks. The remreed switch, with its associated diodes, transistors, and silicon integrated circuits (IC), replaces the ferreed switches, latching relays, and discrete solid-state logic devices used in the ferreed networks.

The remreed switch is similar in many respects to the ferreed. It has two sets of overlapping magnetic reeds, and each pair is hermetically sealed in a glass envelope analogous to the ferreed. The reeds are made of remendur which is a semipermanent magnetic alloy. The remendur reeds are magnetized by current pulses sent through the control windings so that they will hold or latch the sealed contacts in either a closed or open state. In contrast, the ferreed switch requires external remendur plates to provide the latching action. The remreed switch, consequently, provides a magnetic field that is closer to the reed contacts and results in more efficient operations. Also,

**Fig. 7-11.** Remreed Switch Assembly (Courtesy of Bell Telephone Laboratories)

magnetic interaction between adjacent remreed crosspoints is substantially reduced as compared to ferreed crosspoints.

Remreed switches are usually arranged in an 8 by 8 array. This assembly contains 128 crosspoints compared to the 64 crosspoints provided by an equivalent array of ferreed switches. The sealed reed contacts of the remreed crosspoints are held between two printed circuit boards which also contain the diodes and transistors for the electronic control of the remreeds. Figure 7-11 is a photograph of a typical remreed switch package or assembly.

Figure 7-12 shows the major components of a remreed and a schematic drawing of its control windings. The remreed crosspoints are divided into upper and lower halves by a steel shunt plate similar to the construction of the ferreed switch. The magnetization of each half is controlled by a primary and a secondary differential winding; one is in the horizontal pulse path and the other in the vertical pulse path as depicted in Fig. 7-12. These two differential windings are wound on a form which surrounds the glass envelopes containing the reed contacts.

When a current pulse is sent only momentarily through the vertical pulse path, $V_1$ $V_2$, magnetic fields are produced in opposite directions above and below the steel shunt plate. These magnetic fields will magnetize the up-

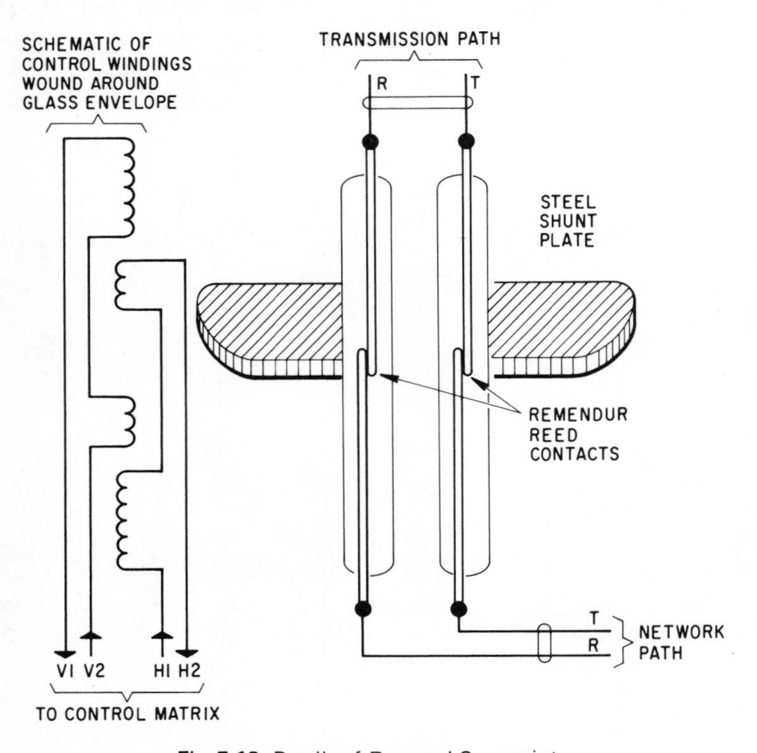


**Fig. 7-12.** Details of Remreed Crosspoint

per and lower parts of the sealed reed contacts in opposite directions, thus the contacts will release. For the same reason, a momentary current pulse only through the horizontal pulse path, $H_1H_2$, likewise will cause the reed contacts to release. When current pulses are sent through both vertical and horizontal pulse paths simultaneously, a magnetic field will be produced in the same direction both above and below the steel shunt plate, and the reed contacts will close.

The crosspoints or contacts of the remreed switch are associated in pairs, as in the case of the ferreed. One set of contacts is used for the tip (T) conductor and the other set for the ring (R) conductor of the two-wire line, trunk, or other network transmission path. The remreed switch, as the ferreed, employs the destructive current mark principle. That is, a current pulse passing through a particular column and row of remreed crosspoints will operate only the specific crosspoint at the intersection. All other crosspoints along the path will release. The operated crosspoint connects the T and R leads associated with the column to the two-wire pair associated with the row. At the same time, crosspoints in other rows and columns, which may have been left connected from previous calls, will be released. Thus, when a call is disconnected or abandoned, the path memory in call store is marked so that the links that were used by that call are idle. This destructive mark ensures that these links will be disconnected from their previously established path when needed for a succeeding call.

## Master Control Center

The operation, administration, and maintenance of electronic central offices, such as the No. 1 ESS with its self-diagnostic features, are handled by a master control center. It is usually centrally located in the central office and serves as the focal point for communications between the electronic switching system and maintenance personnel.

The master control or maintenance center consists of four main equipment units. They are designated (from right to left in Fig. 7-13) the alarm display and control panel, trunk and line test panel, teletypewriter unit, and the automatic message accounting recorders. The memory card writer unit, described in Chapter 5 and Fig. 5-10, sometimes is installed adjacent to the master control center. A drawing of a typical master control center is shown in Fig. 7-13 and a photograph of this equipment appears in Fig. 2-16.

The alarm display and control panel may be considered as the centralized control point of the electronic central office. It contains four sections each with lamps, keys, pushbuttons, and related rotary switches. (The status display lamps in this panel show the status of the system's operations. In case of a trouble condition, maintenance personnel can quickly ascertain the seriousness of the failure by glancing at the lamp indications.) The emergency action section contains keys and lamps to take over system control in case of

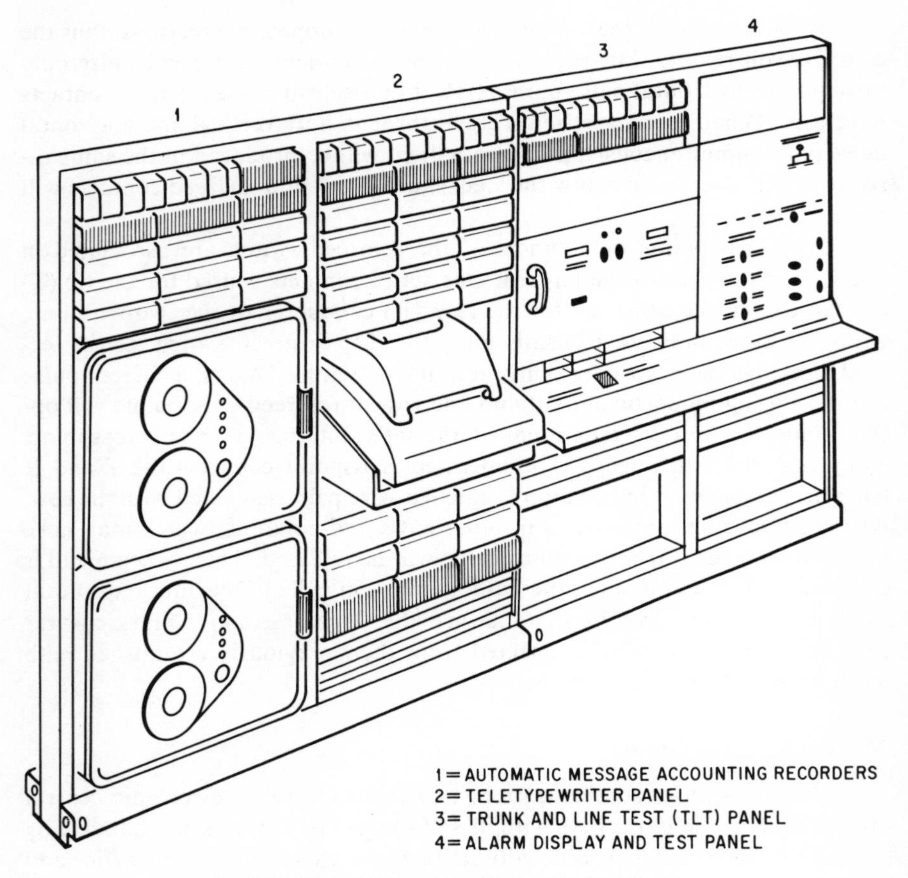


**Fig. 7-13.** Major Units of the Master Control Center

program failures. The communication bus control section has keys and lamps for control of the peripheral bus systems and display of their status. Keys in the program interrupt control section are used to insert program data into the system as may be required in emergencies. Likewise, displays of data from the system can be requested on the program display lamps in the program interrupt control section.

The trunk and line test or TLT panel furnishes the means for testing and removing from service, if necessary, all outgoing trunks, service circuits, and customer lines. Facilities for the disposition of permanent signals, particularly those caused by receivers off-hook, are also included. A master line, associated with a Touch Tone® set, is connected to the line-link network in the same manner as a customer's line. This arrangement is illustrated in Fig. 7-14. The master test line, therefore, can be connected to any trunk or service circuit for test purposes. Consequently, it is not necessary to provide individ-

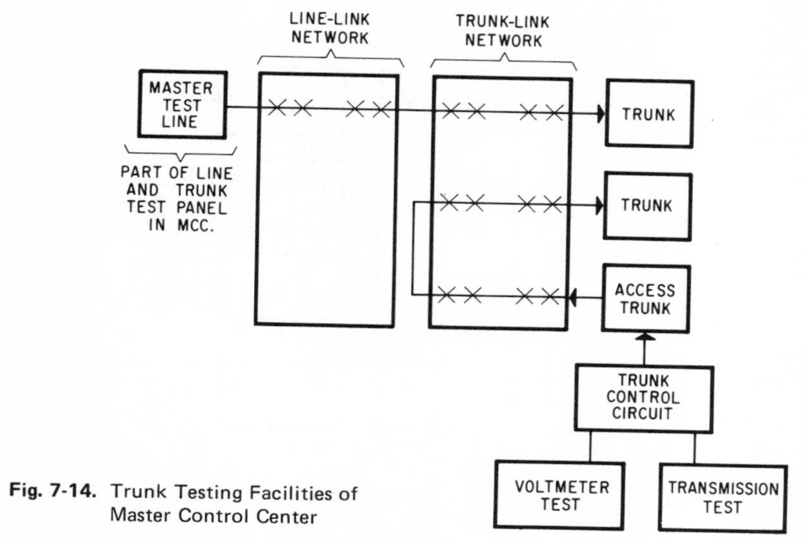


**Fig. 7-14.** Trunk Testing Facilities of Master Control Center

ual test or outgoing test (OGT) jacks for every outgoing trunk as is the case with electromechanical central offices. In addition, any trunk or service circuit can be connected to the voltmeter or transmission test circuits through one of the three access trunks as indicated in Fig. 7-14. Any customer line, likewise, can be reached through the line-link network for connection to the voltmeter test circuit. Figure 7-15 shows the network connections to idle and busy customer lines. The no-test circuit, associated with the test vertical in the junctor switching frame (see Fig. 7-9), can only be activated from the master control center to test an off-hook line.

Teletypewriters are the main means of communication between the electronic switching system and maintenance personnel. In addition to the teletypewriter unit (TTY) provided in the master control center, another one may be located in another part of the central office or at a remote maintenance center if the electronic office will be unattended at times. Also, other teletypewriter channels may be connected with test bureau and traffic centers as illustrated in Fig. 7-16.

Teletype messages typed into the electronic switching system from the master control center indicate the action to be taken. For instance, the system may be asked to trace a call through the network from a specific customer line or trunk. The system may also be ordered to perform a diagnostic test on certain parts. Output messages, printed by the teletypewriter, are either information or action types. Information messages, for example, may pertain to a list of busy trunks, unusual operating conditions or traffic and administrative data requiring no immediate work. Action messages, on the other hand, require immediate or subsequent operations by maintenance personnel depending on the action category. For instance, a major alarm due to a trouble con-

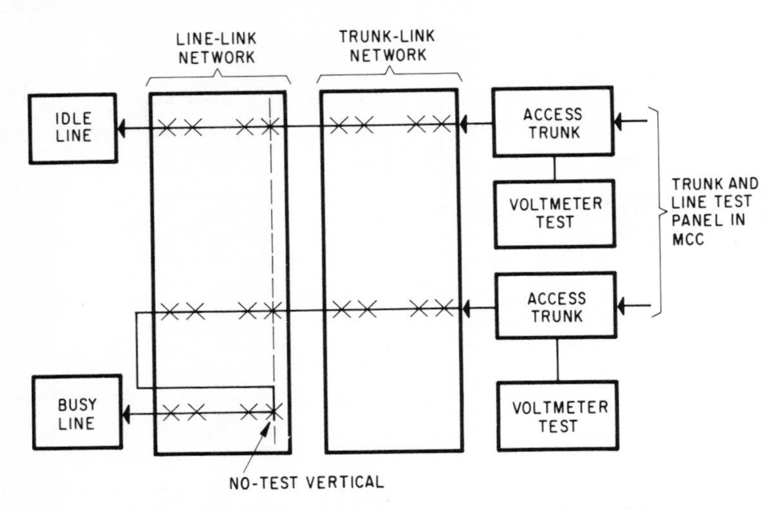


**Fig. 7-15.** Line Testing Connections from Master Control Center

dition in central control requires immediate action. A minor alarm caused by a failure in a service circuit, however, may be deferred for later diagnosis.

The automatic message accounting (AMA) recorder stores on magnetic tapes the charging information pertaining to calling customer lines. These tapes are subsequently transported to a data processing center for use in computing customer bills. One 2,400-foot reel of tape is usually adequate for storing the billing data for about 100,000 calls in one day. A duplicate AMA unit is provided in the master control center to ensure reliability and recording continuity. The sequence followed for recording AMA data on a call is illustrated in Fig. 7-17. If the call processing program determines that AMA data is to be recorded on a particular call, the pertinent information will first be

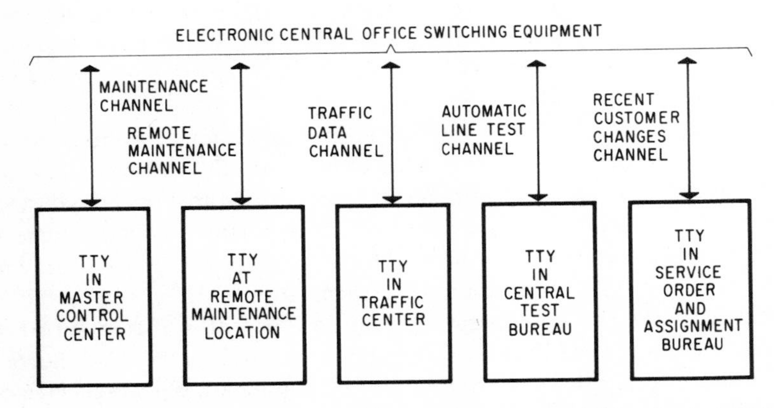


**Fig. 7-16.** Teletypewriter Channels to Master Maintenance and Other Control Centers

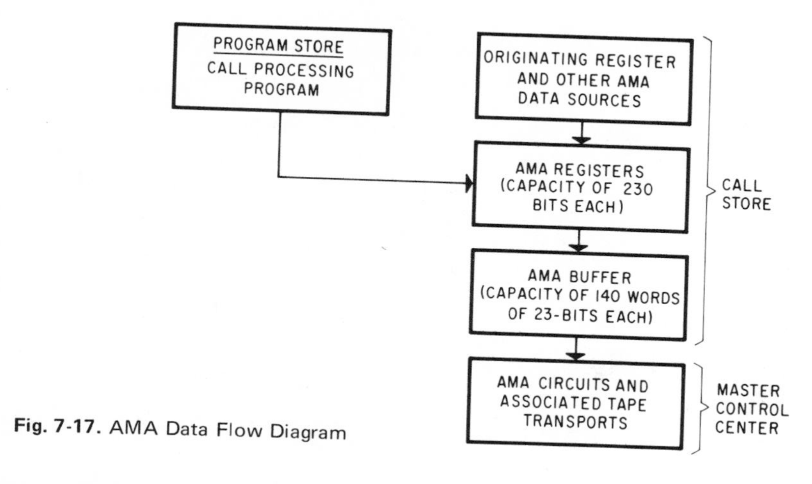


**Fig. 7-17.** AMA Data Flow Diagram

temporarily stored in an AMA register in call store. Upon completion of the call, the related data are assembled in the binary-coded decimal (BCD) format, as explained in Chapter 3 and Table 3-1, and then placed in an AMA buffer in call store. The recording procedure is initiated by an AMA program in program store when the AMA buffer in call store is fully loaded. This AMA program will cause central control to direct that the data be transferred, one word at a time, to the AMA circuit for recording on the magnetic tape. This procedure is shown by the block diagrams in Fig. 7-17.

## Power Plant System

The power plant system for the No. 1 ESS and similar electronic central offices includes two common battery power supplies. One supplies −48 volts and the other furnishes +24 volts to the power distributing frames in the central office. Both battery plants can handle voltage tolerances of plus or minus 10 percent. This design avoids the need for end cells and counter cells that are provided in battery plants of electromechanical central offices in order to maintain stable voltage conditions. The normal input power source is commercial ac power at 208/120 volts, 60 Hz, 3-phase, which employs four-wire transmission. To protect against possible prolonged failures of the commercial ac power, one or more engine-alternator sets are installed.

A simplified block diagram of a typical power system for electronic switching offices appears in Fig. 7-18. The +24 volt and the −48 volt storage batteries have sufficient capacities to handle traffic loads during busy hours in case of an ac power failure. The −48 volt battery, likewise, supplies power to the coin control and the ringing and tone supply power plants. It also has adequate capacity for operating the emergency ac power plant which provides 3-phase, 208/120 volts, 60 Hz power to the ringing plant, announcement machines, teletypewriters, and the AMA magnetic tape recorders in the event of

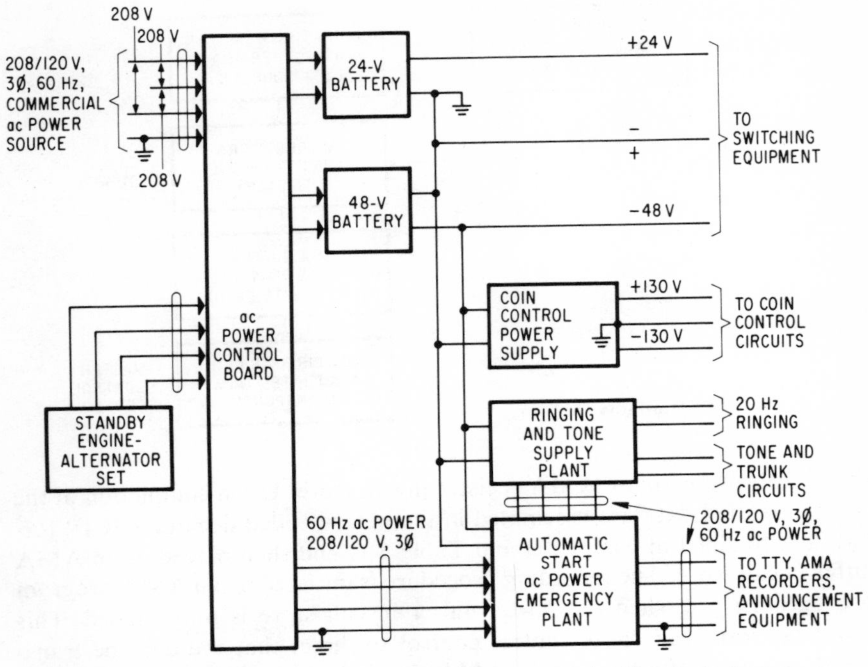


**Fig. 7-18.** Power Systems Including dc Power Supplies for Electronic Central Office

a commercial ac power interruption. This emergency ac power plant starts automatically whenever an ac power failure occurs.

Three power feeders,–48 volts, ground, and +24 volts, as shown in Fig. 7-18, are run from the power room to power distributing frames in the switch room. Each battery feeder is filtered by a 35,000 μF capacitor bank. Distribution of power to the equipment frames is arranged so that no two duplicate circuits are connected to the same power distributing frame. Moreover, to

**Table 7-1.** Designated Tones and Frequencies Used in Electronic Switching Offices

| Tone Designations | Frequencies |
|---|---|
| *Dial tone | 350 Hz plus 440 Hz |
| *Audible ringing tone | 440 Hz plus 480 Hz |
| *High tone | 480 Hz |
| *Low tone | 480 Hz plus 620 Hz |
| Line busy tone | Low tone at 60 IPM |
| Path busy tone | Low tone at 120 IPM |
| Ringing current | 20 Hz |

*Fundamental tones
IPM = Interruptions per minute

minimize electrical noise, all frames and cable racks are insulated from the building at the time of installation. Power plant ground returns are similarly insulated from their power plant frames.

The 20 Hz ringing voltage is produced by a duplicate set of ringing generators. The audible ringing tone and all other tones are derived from four fundamental tones, generated by precision transistor oscillators, and amplified. These fundamental tone signals are 350 Hz, 440 Hz, 480 Hz, and 620 Hz. All utilized tones, except the high tone, consist of a combination of two fundamental tones. The actual output of each transistor oscillator is a square wave. However, it is converted by bandpass filters to a sine wave, with a harmonic level at least 60 dB down from the basic frequency, in order to produce the required precision tone. Table 7-1 lists the various tone designations and their precise frequencies as employed in electronic central offices.

## Questions

1. Why is it necessary to employ switching devices that physically open unused parallel paths in electronic central offices?
2. How many switching stages are provided in the switching network? Where are they located?
3. What are the main components of the crosspoint ferreed switch? Where is it employed?
4. What are the main functions of the bipolar ferreed switch?
5. What are the major elements of the switching network and what frames comprise each element? What traffic concentration ratios are normally provided and where?
6. What links and frames are in the transmission path from the calling to the called line on an intraoffice call? On an interoffice call?
7. What binary digit readouts from the ferrods in the junctor circuit indicate that both parties are connected with each other? What binary readouts indicate that both parties have disconnected? Why?
8. How is the remreed switch similar to the ferreed? What are the main advantages of the remreed switch?
9. What are the major elements of the master control center? What types of messages are printed by the teletypewriter? Give an example of each type of message.
10. Which batteries make up the dc power system in an electronic central office? What are the fundamental tone signals from which all other tones are derived? What frequencies are used for dial tone and for the line busy tone?

# 8
# ELECTRONIC TELEPHONE SWITCHING DEVELOPMENTS

## Improvements in Electronic Switching Technology

The basic concepts of electronic telephone switching systems are exemplified in the foregoing descriptions of the No. 1 ESS. It may be considered as a high-speed processor which operates with a stored program of instructions and translation information. This coalition serves to control the call-processing operations of the system on a time-shared basis. The stored program or *software,* as it is commonly called, makes possible the great flexibility of the No. 1 ESS and other similar types of electronic switching systems.

Electronic telephone switching systems are often regarded as a type of computer. The stored program or memory phase of a general purpose computer and the program store in the No. 1 ESS are similar, but their dependability requirements differ considerably. For instance, a temporary stoppage of the computer may delay the output results which would usually be an inconvenience only to the user. Any stoppage of the electronic switching system, however, would halt telephone service in the central office which would be a disastrous situation. On the other hand, a data processing error in the No. 1 ESS may result only in a wrong number on the particular call being processed. In a computer, a processing error cannot be tolerated because it may cause, among other things, an overcharge in an account. It is seen, therefore, that the dependability requirements of the No. 1 ESS and other electronic switching systems are very high. This necessity has resulted in the duplication of many subsystems and peripheral units in electronic switching systems. This equipment duplication is not usually required in computers.

Recent technological developments in solid-state devices, particularly large scale integrated circuits, are expected to reduce substantially the cost and size of electronic telephone switching systems. For example, about 5,000 circuit packs are employed in the No. 1 ESS, each containing a number of discrete components including transistors, diodes, resistors, and capacitors.

One integrated circuit, about the size of a postage stamp, can contain many hundreds of transistors, diodes, and resistors. Thus, the overall cost and space savings for an electronic central office would be considerable.

Moreover, improved integrated circuit techniques can also be applied to the memory units of electronic switching systems. A new toll electronic switching system, employing integrated circuits for the program and call store memory units, has been developed by the Bell Laboratories and designated the No. 3 ESS. It may replace the present No. 4 Toll Crossbar offices. Likewise, the trend is to replace the ferrite cores and the aluminum cards with their bar magnets and associated twistor wires used in the No. 1 ESS, with semiconductor memory units utilizing integrated circuits. With this improvement, the same number of binary words in program store, for example, could be handled by one tray of equipment instead of the four equipment bays currently needed.

The basic design of the Bell System's No. 1 ESS has led other laboratories and telephone organizations to develop electronic switching systems predicated on the use of the stored program and call processor techniques. Table 8-1 lists some of these systems and compares them with the No. 1 ESS. Since it is not within the scope of this book to describe all of the other electronic telephone switching systems, only succinct explanations will be presented of the Bell System's No. 2 ESS and the No. 1 EAX of the GTE Automatic Electric Company.

**Table 8-1.** Major Electronic Telephone Switching Systems

| System Designation | Manufacturer | Maximum Terminals | Maximum Busy-hour Calls |
|---|---|---|---|
| No. 1 ESS | Bell System | 96,000 | 105,000 |
| No. 2 ESS | Bell System | 10,000 | 19,000 |
| No. 1 EAX | GTE Automatic Electric Co. | 30,000 | 79,000 |
| SP-1 | Bell-Northern Electric Co. (Canada) | 20,000 | 36,000 |
| NX-1E | North Electric Co. | 20,000 | 68,000 |
| D-10 | NTT (Japan) | 65,000 | 90,000 |
| METACONTA L | ITT System | 64,000 | 100,000 |

## No. 2 ESS Design Precepts

The No. 1 ESS central office described in this book has the capability of serving many tens of thousands of telephone lines with heavy traffic requirements in large cities and metropolitan areas. It would not, however, be economical to utilize the No. 1 ESS for the telephone switching requirements of smaller localities and suburban areas. Consequently, the No. 2 ESS was designed by the Bell System primarily for use in nonmetropolitan areas and where an electronic central office would serve approximately 1,000 to 10,000 customer lines. Furthermore, the No. 2 ESS has been designed to be largely unattended, and to be operated by remote control in most maintenance and test functions.

The service features offered by the No. 2 ESS are essentially the same as those provided by the No. 1 ESS as described in Chapter 2. The No. 2 ESS, moreover, can provide four-party and eight-party line service which is still needed in some suburban and rural areas. Its traffic design capacity is for 19,000 busy-hour calls, and can handle up to 10,000 lines in small urban localities and suburban areas. In case of large unanticipated growth, two No. 2 ESS offices may be combined into a dual central office. For this arrangement, data links would be provided for communication between the central office units of each office, and their switching networks would form a common network for all trunks and service circuits of both offices.

Many of the frames and apparatus developed for the No. 1 ESS are also utilized in the No. 2 ESS. For instance, the peripheral equipment such as

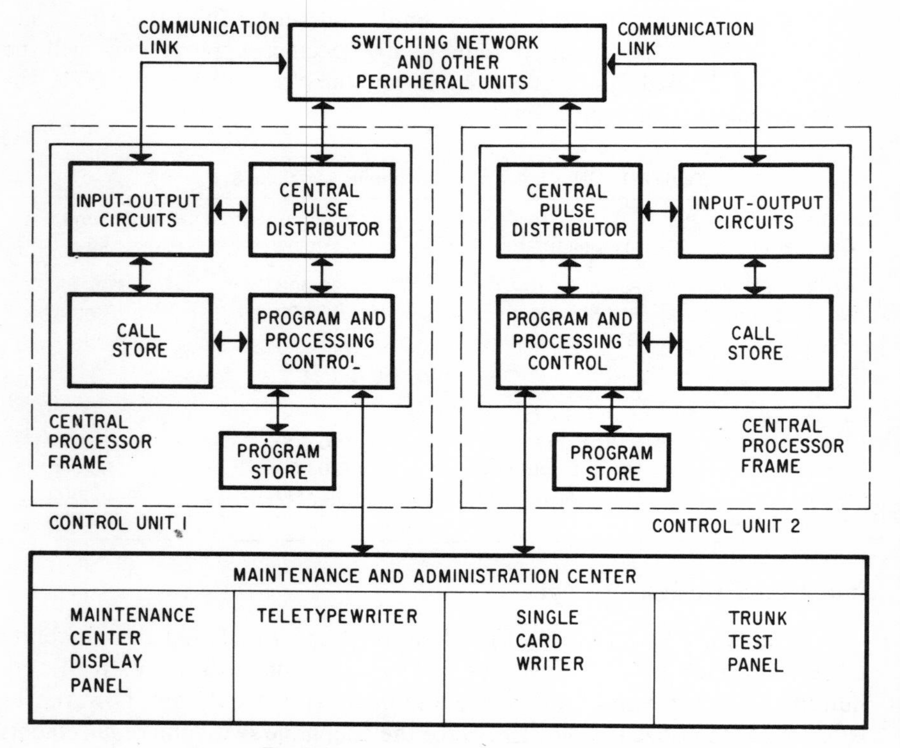


**Fig. 8-1.** Control Entities of No. 2 ESS

ferreed switches, ferrod scanners, and the trunks and service circuits of the No. 1 ESS are included in the No. 2 ESS. Likewise, the program and call stores in the No. 2 ESS employ the same memory devices and operate in a manner equivalent to their counterparts in the No. 1 ESS. Consequently, only the apparatus and functions of the No. 2 ESS not previously explained will be specifically described in this chapter. (The reader is referred to other chapters for applicable details concerning equipments and functions common to both No. 1 ESS and No. 2 ESS central offices.)

## Basic Elements of No. 2 ESS

The major components of the No. 2 ESS (see Fig. 8-1) are arranged in a control complex consisting of two control entities, each associated with a common maintenance and administration center. The control entity consists of a central processor frame, and one to four program store frames. A supplementary call store frame may be added if required. The central processor frame provides the control functions for operating the No. 2 ESS. It contains the call processing or program control circuits which may be considered equivalent to central control, the call stores, a central pulse distributor, and input-output control circuits in the No. 1 ESS. The two independent control units can be switched as a complete entity rather than as individual units, as in the No. 1 ESS.

Consequently, the number of frames and equipment quantities required are considerably reduced for the No. 2 ESS. For example, the major equipment frames are the central processor, program store, line-trunk switching, network control junctor switching, junctor grouping, and the maintenance center with its trunk test frame. Other important peripheral equipment frames are the master scanner, supplementary central pulse distributor, automatic message accounting, and the various trunk frames, including the universal trunk and junctor, and the miscellaneous trunk frames. These major frames and their correlations are shown in the block diagram of Fig. 8-2.

The two control units or entities, illustrated in Fig. 8-2, synchronize with each other. Control unit 1 normally is "on-line" or in command of the call processing functions. The operations of both control entities are matched at the maintenance and administration center. Thus, in case of any failure, control unit 2 instantaneously takes over operations. The call store in each control unit is shared by both the program and processing control and the input-output control circuit. Each control unit connects to the switching network and other peripheral units over a peripheral bus system or communication link. These buses, as later explained in Fig. 8-5, provide access to the duplicate controllers of the various peripheral frames, such as the master scanner, line-trunk switching, network control junctor switching, and the trunk frames.

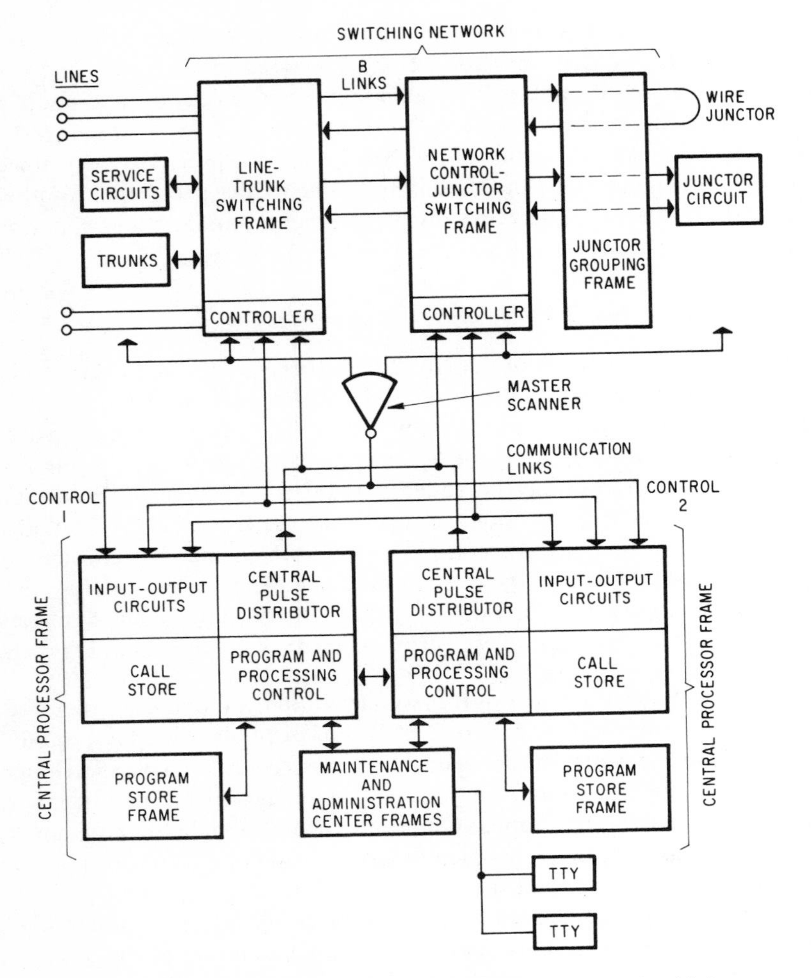


**Fig. 8-2.** Major Frames of No. 2 ESS and Communication Links

## No. 2 ESS Central Processor Functions

The central processor frame in the No. 2 ESS, a photograph of which appears in Fig. 8-3, performs the major call processing functions. Its main components, as depicted in Fig. 8-2, are the program and processing control unit, call store, input-output control circuits, and the central pulse distributor. The program and processing control unit directs the processing of calls in a manner equivalent to central control in the No. 1 ESS. It also detects and diagnoses equipment malfunctions with the aid of the maintenance center.

The central processor contains logic circuits to write and read out up to 32,768 binary words (each of 16 bits) of information that can be temporarily

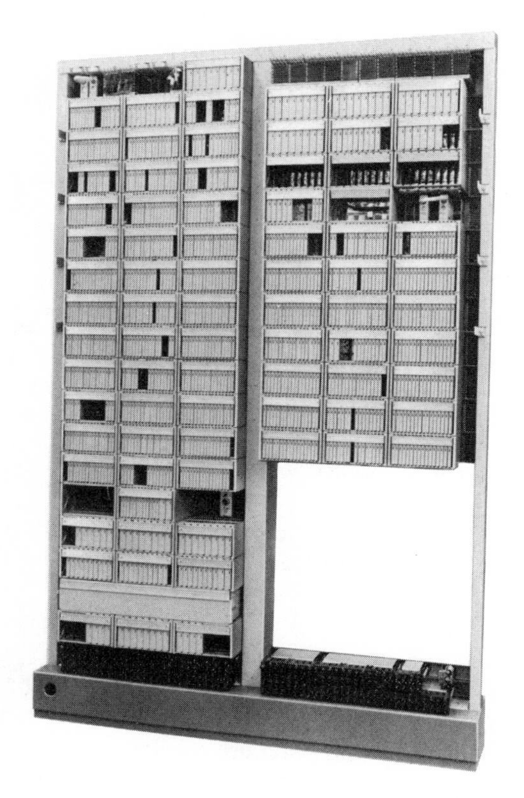

**Fig. 8-3.** Central Processor of No. 2 ESS (Courtesy of Bell Telephone Laboratories)

stored in up to four call stores. Two of these call store units are installed in the central processor frame, and the other two, when required, are mounted in a supplementary call store frame. The call store employs the same one-inch square ferrite sheet memory device used in the No. 1 ESS call store, which is described in Chapter 5.

Call store operations are similar in many respects to those in the No. 1 ESS as described in Chapter 5, particularly Figs. 5-13 through 5-15. Temporary storage for a random-access memory of 8,196 binary words of 16 bits each is provided in a call store. There are no registers in call store because direct wire connections are used to receive address and input data from pertinent registers in the program and processing control circuit. Likewise, output data from call store is sent by dc pulses directly to registers in either the input-output circuit or to the program and processing control circuit, depending upon the readout instructions. These register relationships and other elements of call store are depicted in Fig. 8-4.

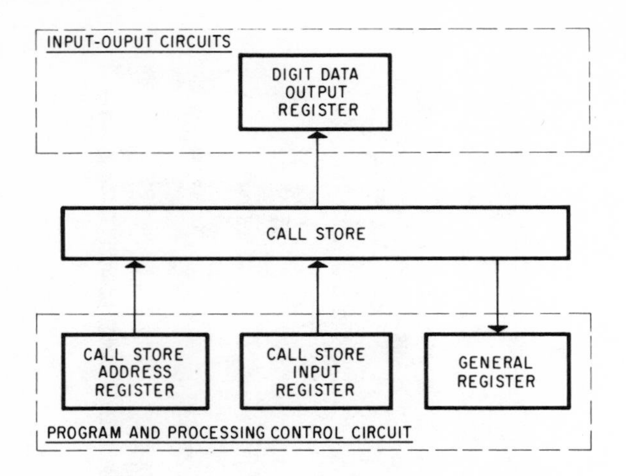


**Fig. 8-4.** Call Store Operation Registers in No. 2 ESS

The program and processing control and other central processor units employ large numbers of binary logic circuits in a similar fashion to central control in the No. 1 ESS. The basic logic circuit in the No. 2 ESS, however, is the transistor-resistor NOR gate. A schematic drawing of this logic gate with approximate resistor values is illustrated in Fig. 8-5. The gate has two inputs and a maximum of three outputs or fanouts. NOR gate operating principles are explained in Chapter 3 and illustrated in Figs. 3-8 and 3-9. In the No. 2 ESS, the NOR logic and similar gates are constructed with high-speed silicon transistors and special thin film resistors. Many variations of the basic NOR logic gate have been developed for the different operational requirements of the central processor. Logic gates may have several inputs and as many as ten or more fanouts. A schematic of a typical high fanout transistor-resistor NOR gate is shown in Fig. 8-6. Logic gates also form the binary counter-shift registers that are employed in units of the central processor frame.

The input-output control circuits perform the interface functions between the program and call processing control unit and the peripheral units. Moreover, the input-output control circuits act as part of the program and call processing control unit by providing the necessary facilities for communi-

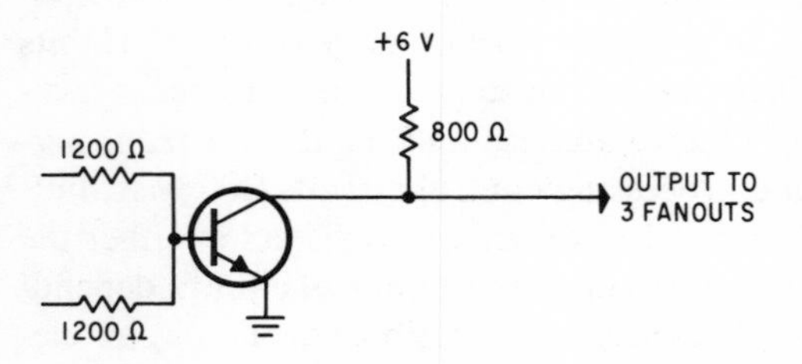


**Fig. 8-5.** Basic Transistor-Resistor NOR Logic Gates in No. 2 ESS

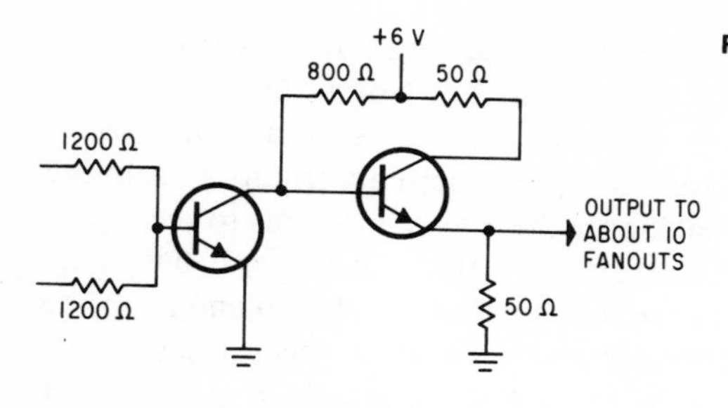


**Fig. 8-6.** High Fanout Transistor-Resistor NOR Logic Gate in No. 2 ESS

cating with peripheral units under control of relevant program instructions. These circuits also provide wired logic for independently performing parts of certain tasks, such as digit receiving, digit and data transmission, and line scanning. To conserve program time, wired logic is used for many simple and highly repetitive tasks, instead of using instructions stored in program store.

The central pulse distributor (CPD) located in the central processor frame, as shown in Fig. 8-2, provides the means of steering control pulses to

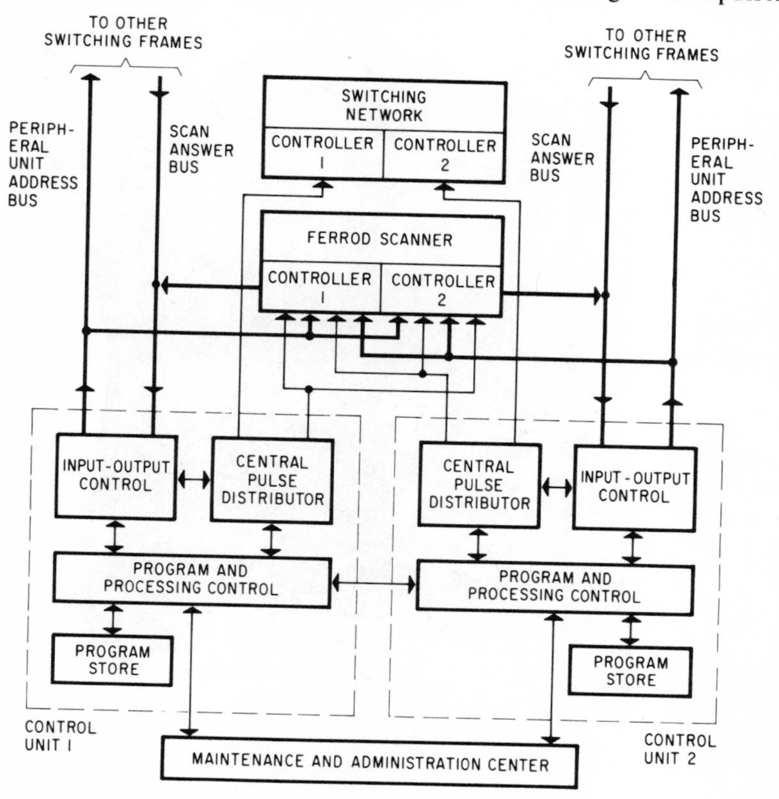


**Fig. 8-7.** No. 2 ESS Peripheral Connections and Bus System

one of 512 outputs of the peripheral units. The central pulse distributor has electrical characteristics similar to the ones in the No. 1 ESS. Its output pulses are used to select and enable a peripheral unit, such as a line-trunk switching frame, or a trunk or junctor circuit, to receive information from the address bus, and to transmit data back over the scan answer bus to the input-output control circuit. Figure 8-7 depicts the various peripheral bus functions. Balanced twisted pairs, which are transformer-coupled to and from the associated circuits, are used for the peripheral bus system. Operations can be compared to the equivalent buses of the No. 1 ESS as explained in Chapter 4 and shown in Fig. 4-4.

## Program Store Operation in No. 2 ESS

The semipermanent information and memory data for control units 1 and 2 in the No. 2 ESS are furnished by their respective program stores. Program store memory consists of the random-access type but is designed for

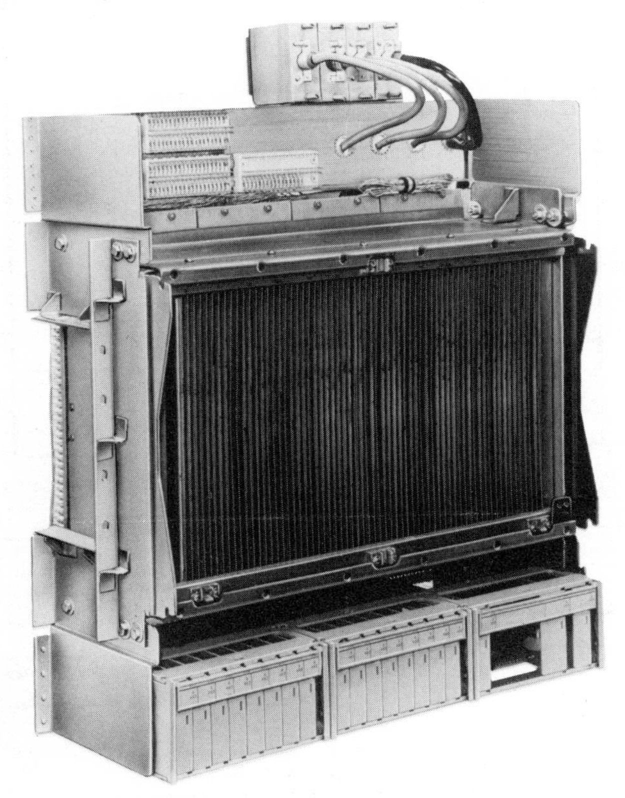

**Fig. 8-8.** Program Store Unit of No. 2 ESS (Courtesy of Bell Telephone Laboratories)

readout only. It operates within an access time of 2.5 microseconds and a 6-microsecond cycle time. A total of four program store frames may be equipped, each with 16,384 binary words (22 bits per word), or a total of 65,536 binary words.

The memory device employed in the No. 2 ESS program store is the same permanent magnet twistor-wire type with aluminum cards containing miniature magnets, that are used in the No. 1 ESS. This memory mechanism, shown in Fig. 8-8, is explained in Chapter 5 and illustrated in Figs. 5-3, 5-5, 5-6, and 5-8. Information in program store may be changed by withdrawing the aluminum cards from the twistor module and then changing the state of pertinent miniature magnets just as in the case of the No. 1 ESS.

A program store in the No. 2 ESS contains four memory units and various associated equipment control circuits as shown in Fig. 8-9. Each unit includes a twistor wire memory module and its related address decoder and access circuits. The program store connects to the program and processing control unit in the central processor frame over the address and answer ac bus system. Address data for locating the particular memory unit, and the desired program instructions or translator information in it, is received over the address bus by the 18-bit address register. The information readout is first amplified and then transmitted back to the control unit over the answer bus.

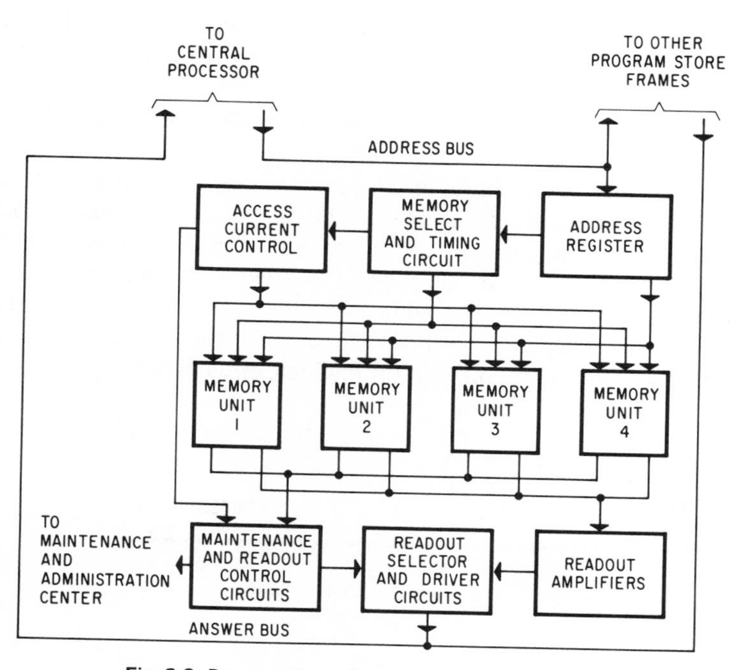


**Fig. 8-9.** Program Store Operations in No. 2 ESS

To detect circuit malfunctions quickly,program store has checking circuits which verify correct operations on every readout cycle. This verification is accomplished by checking the access circuit's current flow in the memory module, and seeing that the correct current path has been selected. In case of a failure, an error signal is sent to the program and processing control circuit and to the maintenance center. Also, as in call store, dc control signals are sent by the maintenance center to program store to simulate different types of malfunctions that the check circuits are designed to detect. In this way, the integrity of the check circuits is verified.

The basic 22-bit instruction words in program store memory consist of two types. One type contains a 5-bit operation code, a 16-bit address, and one transfer check bit. There are relatively few instructions of this first type of instruction word and they are used mainly for program transfers. The other type, which contains most of the instructions, consists of two 10-bit instructions, each with a 5-bit operation code and a 5-bit address, one transfer check bit, and a parity check bit. This instruction format is shown in Fig. 8-10.

The central processor can decode and execute only one instruction at a time, just as central control does in the No. 1 ESS. Consequently, the 22-bit instruction format, read out from program store, is first temporarily stored in an output buffer until the preceding instruction has been completed. The current instruction is then gated to the program store output register for decoding and completion. Next, the program address register is advanced by 1 so that the following sequential binary word will be read out from program store. In this manner, the succeeding program word can be read while the current instruction in the program store output register is being executed, thereby reducing operating time.

Instructions placed in the program store output register are translated by a two-stage decoding arrangement. This consists of the main decoder and one of eight auxiliary decoders as illustrated in Fig. 8-11. The main decoder selects a particular auxiliary decoder. The selected or enabled auxiliary decoder then translates the 5-bit address to one out of thirty-two connections that lead to the desired program instruction. Outputs of the main decoder also connect to the command and timing logic circuits in order to control the

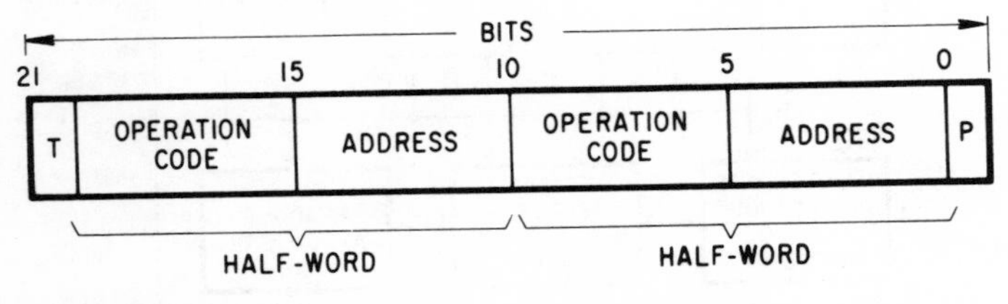


**Fig. 8-10.** Program Store Instruction Format Used in No. 2 ESS

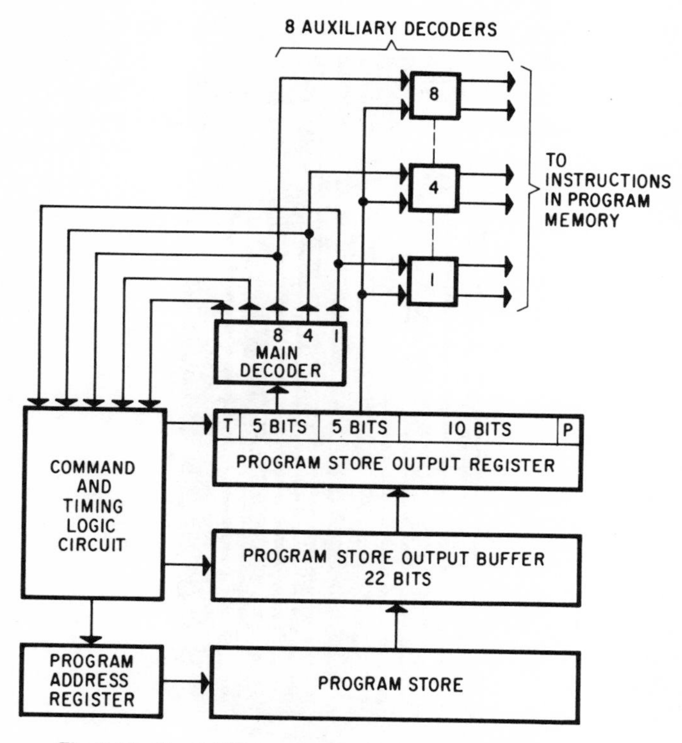


**Fig. 8-11.** No. 2 ESS Program Store Decoding Arrangement

transmission of instructions from program store to the decoders, and to advance the program address register by 1 as previously explained. Subsequent program store operations are similar in many respects to the program store functions of the No. 1 ESS as described in Chapter 6.

## The No. 2 ESS Switching Network

The No. 2 ESS switching network, shown in Fig. 8-2, is of the space division type similar to the No. 1 ESS. It consists of from one to four line-trunk switching frames (see Fig. 8-12), each associated with a network control junctor switching frame (see Fig. 8-13). The normal capacity of the No. 2 ESS is fifteen line-trunk networks, which can provide up to 30,720 terminals, assuming a 4 to 1 concentration ratio. Eight stages of ferreed switches, which form a four-stage folded network, are used for interconnecting the two-wire (tip and ring) metallic paths in the switching network.

Unlike the No. 1 ESS switching network, customer lines, trunks, and service circuits in the No. 2 ESS are assigned to terminals on the same side of line-trunk switching frames. These terminals may be interconnected by

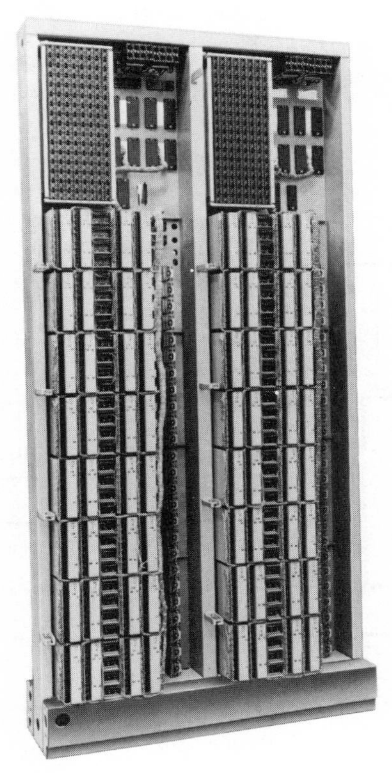

Fig. 8-12. Line-Trunk Switching Frame in No. 2 ESS (Courtesy of Bell Telephone Laboratories)

means of the aforementioned eight switching stages and through wire junctors or junctor circuits, depending upon the type of call. The first two and the last two switching stages are situated on the line-trunk switching frame. The third, fourth, fifth, and sixth switching stages are located on the network control junctor switching frame. This eight-stage switching arrangement is illustrated in Fig. 8-14.

The line-trunk network, under control of the central processor, provides the necessary paths to establish intraoffice and interoffice calls, as well as connections to service circuits and to the maintenance center. For instance, an interoffice call is shown in Fig. 8-14 by the paths in the line-trunk network, from calling line A through switching stages 1-4, the wire junctor, and back through switching stages 5-8 to outgoing trunk B. Each line in this drawing represents a tip and ring metallic pair. A line ferrod and scanner circuit is connected to line A through the cutoff contacts, designated C, of its associated ferreed switch.

**Fig. 8-13.** Network Control Junctor Switching Frame in No. 2 ESS (Courtesy of Bell Telephone Laboratories)

A ferreed switch, designated F on the network control junctor switching frame, is used to connected a B link to the test vertical on this frame. This test vertical may be connected, under control of the central processor, to either the false cross or ground (FCG) test circuit or to the no-test circuit. The FCG test circuit checks the assigned network paths for false crosses, grounds, or any foreign potentials. The no-test circuit enables maintenance personnel to check for false busy conditions on lines and trunks. These operations are analogous to those performed in the No. 1 ESS as explained in Chapter 7 and Fig. 7-7.

An example of an intraoffice call is also depicted in Fig. 8-14. It is represented by the paths from calling line C through the four switching stages, numbered 1 to 4, in the line-trunk network to the junctor circuit, and from the junctor circuit back through switching stages, numbered 5 to 8, to called line D. A junctor circuit is employed on all intraoffice calls in order to provide talking battery and supervision on these calls. Trunk circuits provide these fa-

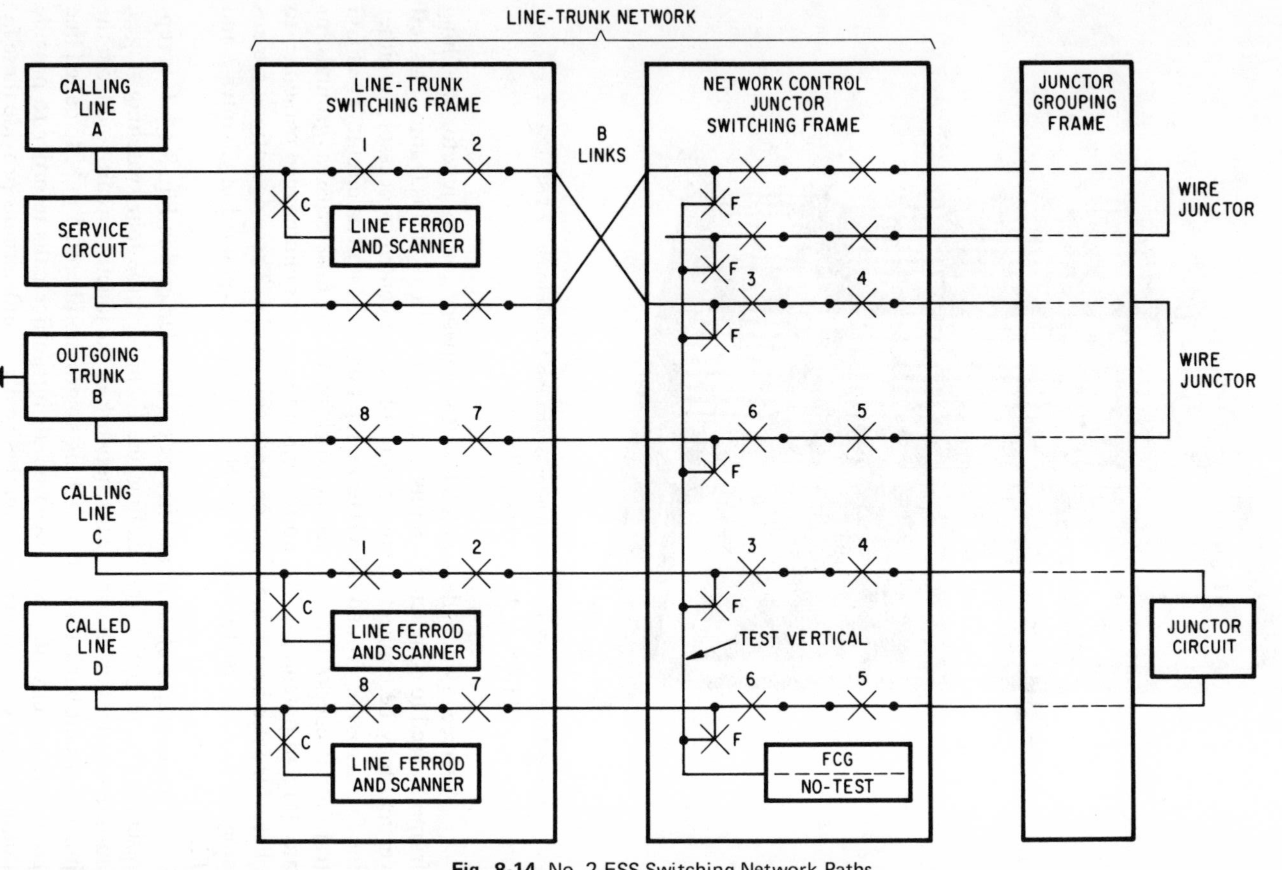


**Fig. 8-14.** No. 2 ESS Switching Network Paths

cilities for interoffice calls so that only wire junctors (metallic connections) need be utilized for such calls.

The switching network is controlled by a pair of controllers in a similar fashion as in the No. 1 ESS. Each controller has access to the entire network. Both network controllers may operate simultaneously, provided that the two paths being connected are in different concentrator groups and not in the same switching stages. Network controllers receive path information and other data from the central processor. The data is transmitted, in the form of bipolar binary pulses, by the central pulse distributor as indicated in Fig. 8-7. Internal check circuits monitor operations of the network controllers to preclude malfunctions. Three scan points are associated with each controller for this purpose. These scanner points, which are terminated at the master scanner shown in Fig. 8-2, indicate the state of the network controller at any time.

## Maintenance and Administration Center in No. 2 ESS

The major elements of the maintenance and administration center in the No. 2 ESS are shown in Fig. 8-15 as follows: the display panel, trunk test panel, teletypewriter, and the single card writer. The separate trunk test frame is not shown in this photograph. Interconnections of the maintenance and administration center with control units of the No. 2 ESS are portrayed in the block diagrams of Figs. 8-1 and 8-2.

The maintenance center frame contains the equipment for the manual control of the No. 2 ESS. It also includes display panel controls, teletypewriter, and associated interface units for communicating with maintenance personnel. The maintenance center serves primarily as a system maintenance tool to ensure continuous and efficient performance of the No. 2 ESS. In this respect, its principal functions are to monitor the operational status of the system, to provide for manual control of operations in the event of failure of diagnostic procedures or in emergency situations, and to provide for routine test and related control operations.

In addition to the teletypewriter furnished in the maintenance center frame, up to a total of eight teletypewriter channels can be handled by the No. 2 ESS. Each such channel includes a tip and ring loop control circuit and connections to a data set. Teletypewriter channels may be established to a remote maintenance center, test bureau, traffic center, and the assignment bureau, just as explained in Chapter 7 and Fig. 7-16 for the No. 1 ESS.

The trunk test frame, usually installed adjacent to the maintenance center frame, is used to make operational and transmission tests on trunks and service circuits. This frame is equipped to make continuity and other tests on trunks and customer lines. The test equipment for this purpose includes transmission and noise measuring sets, impulse counter, peak-average ratio meter receiver, and a voice-frequency oscillator. Complete functional testing of cus-

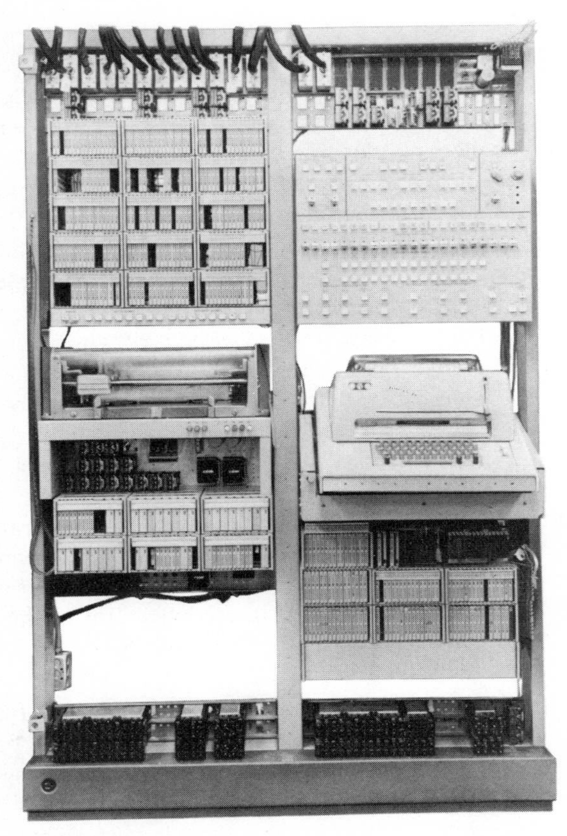

**Fig. 8-15.** Maintenance and Administration Center of No. 2 ESS (Courtesy of Bell Telephone Laboratories)

tomer lines is usually accomplished by separate test facilities, such as the No. 3 local test unit.

The maintenance center frame may also be equipped with a single card writer, as shown on the left side of the frame in Fig. 8-15. The card writer consists of a writing head and means for moving the head over an aluminum memory card. The card writer functions in a manner similar to the card writing equipment installed in the No. 1 ESS and described in Chapter 5.

## No. 2 ESS Power Plant

The design of the power plant in the No. 2 ESS is very much the same as that described for the No. 1 ESS in Chapter 7, except that its capacity is substantially less. Moreover, the ac input source is the usual 120 volt, 60 Hz, single phase power. In addition, two regulated +6.7 volt supplies are derived from the −48 volt battery. The +6.7 volt supplies are used for the logic units in the control complex of the No. 2 ESS.

Three power feeders, carrying −48 volts, ground, and +24 volts, run from the power plant to a power distributing frame. Battery feeders from this frame supply the individual equipment frames in the No. 2 ESS. Also, +130 volts and −130 volts are derived from the −48 volt supply by dc-to-dc solid-state converters for coin control and other requirements. The layout of the power system generally follows that of the No. 1 ESS as depicted in Fig. 7-18.

The 20 Hz ringing voltage and the various tone frequencies are furnished by the ringing and tone frame. This frame provides duplicate 20 Hz ringing generators and the same frequencies as used in the No. 1 ESS for dial tone, audible ringing, busy-tone, and other tone frequencies as shown in Table 7-17 for the No. 1 ESS. Both ac, dc, and superimposed ringing are provided in the No. 2 ESS, for use on four- and eight-party lines.

## No. 1 EAX Electronic Switching System

The No. 1 Electronic Automatic Exchange (EAX) is a stored program and space division type of electronic switching system developed by GTE Automatic Electric Laboratories. It is designed to function as a Class 5 (or end central office), and as a combined Class 4 and 5 (toll and end central office). Generally, the No. 1 EAX would serve large urban localities. The capacity of the No. 1 EAX is 30,000 lines with medium traffic and 64,000 directory numbers; it can grow up to 44,000 lines. Its traffic handling capability is 79,000 busy-hour calls. The types of customer services and other features provided by the No. 1 EAX are essentially the same as those furnished by the No. 1 ESS and described in Chapter 2.

The No. 1 EAX system may be divided into three major groups: the switching network, the common control subsystems, and the maintenance and control center. The switching network includes the line group, trunk register group, and the selector group frames. The common control subsystems are made up of the originating marker, terminating marker, register-senders, and the data processor unit. The maintenance control center (MCC) illustrated in Fig. 8-16 is the focal point for monitoring system operations. It serves as the interface between maintenance personnel and the subsystems of the No. 1 EAX. The maintenance control center also furnishes visual indications of traffic operations and the necessary controls for initiating test call routines and test programs. The teletypewriter provides printouts of maintenance information and related data.

The major groups and associated elements of the No. 1 EAX and their correlations are illustrated in the block diagram of Fig. 8-17. Each line group frame provides for 1,000 customer lines through four stages of switching

**Fig. 8-16.** No. 1 EAX Maintenance Control Center (Courtesy of GTE Automatic Electric Co.)

(used for different purposes), designated A, B, C, and R. Switching stages A and B are used for originating calls, such as from customer line A. These switches select a path from the originating line to a spare originating junctor. Switching stage R connects the selected originating junctor to one of twenty originating register junctors in order to reach a register-sender for subsequent receiving and sending of dialed information. An incoming call would be routed through a terminating junctor to the A, B, and C switching stages in the line group frame in order to be connected to the called party, such as from customer line B. The terminating junctors provide ringing control, talking battery for both parties, and call supervision. Both the originating and terminating junctors remain in the transmission path for the duration of a call.

Three switching stages are provided in the selector group network as shown in Fig. 8-17. They provide interconnections between originating junctors and outgoing trunks, as well as terminating junctors, and between incoming trunks and terminating junctors or outgoing trunks. The trunk register group is equipped with two switching stages. It furnishes a temporary signaling path between 200 incoming trunks and forty incoming register junctors per group. The trunk register group and its associated register junctor provide access to an idle register-sender. These equipment units are utilized

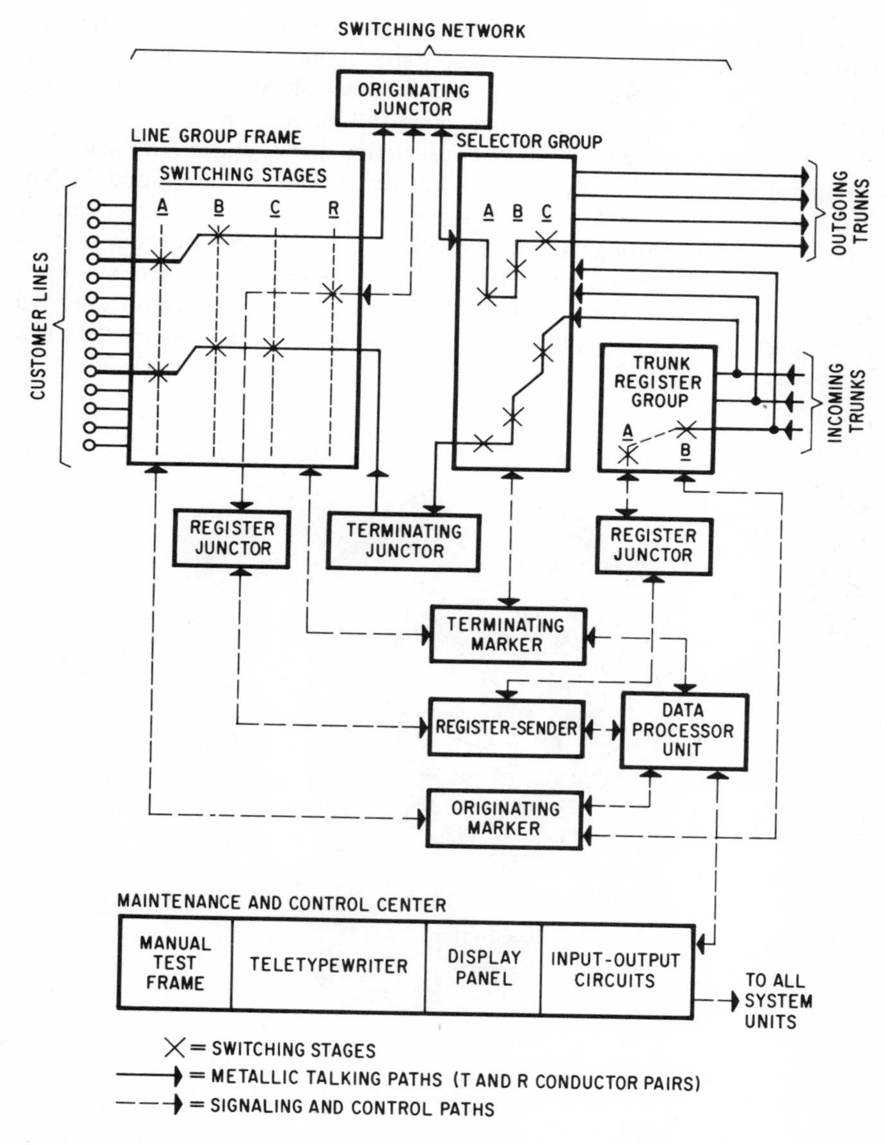


**Fig. 8-17.** Correlations of Major Elements of No. 1 EAX

only during the establishment of an incoming call in order to record the called line number in the register-sender for subsequent processing.

The actual operation of the crosspoint connections in the switching network are performed by *correeds*. The correed consists of two reedcapsules surrounded by a coil winding. Each reedcapsule is composed of two magnetic reeds equipped with gold plated contacts that are hermetically sealed in a glass enclosure or capsule. The reed contacts in one reedcapsule are used to switch the tip conductors, and the reed contacts in the other reedcapsule switch the ring conductors of the particular talking path involved.

The correed is comparable in construction and operation to the ferreed crosspoint switch used in the switching network of the No. 1 ESS and the No. 2 ESS. Unlike the ferreed, however, it is necessary that a small direct current continuously flow through the coil winding of the correed in order that its reed contacts can operate and remain closed. These reed contacts will open when the current flow stops A number of correeds are mounted on a printed circuit card to form the matrix used for a switching stage. Figure 8-18 is a photograph of a typical correed matrix.

**Fig. 8-18.** Correed Matrix Used in No. 1 EAX (Courtesy of GTE Automatic Electric Co.)

## Call Control Operations in No. 1 EAX

The register-sender is a time shared multiplex and common control unit. It has wired logic and memory facilities for receiving and storing dialed digits from originating customer lines, as well as the directory number of the called line on incoming calls. Approximately 390 simultaneous originating and terminating calls can be handled. In addition, the register-sender generates pulses for transmitting outgoing call data via outgoing trunks. A temporary ferrite-core memory is utilized by the register-sender for this purpose in much the same manner as call store in the No. 2 ESS. Similarly, the common logic and memory units in the register-sender are in duplicate and are operated in synchronism to provide instantaneous error and fault detection facilities.

The originating marker detects originating calls from customer lines in the line group frames, and terminating calls from incoming trunks in the trunk register group. It controls path selection between the line group and the originating junctor, and between the originating junctor and the register junctor as indicated in Fig. 8-17. The originating marker also controls the selection of paths between incoming trunks connected to the trunk register group and register junctors. These functions may be likened to those of the originating marker in the No. 1 Crossbar office as explained in Chapter 1 and Fig. 1-10. The originating marker also provides the data processor unit with the identities of the calling line and the originating and register junctors involved, and executes the orders received from the data processor unit for completing the originating call process.

The terminating marker directs the establishment of paths for all calls through the selector group. It also sets up paths through the line group to the called line as represented in Fig. 8-17 by the paths to called line B. The terminating marker functions under the control of the data processor unit. For instance, the terminating marker selects idle paths, either randomly or sequentially, from the specified group, as directed by the stored program in the data processor unit.

Both the originating and terminating markers are installed in duplicate and both operate in pairs. Each can process separate calls simultaneously unless a fault should occur. In this event, the malfunctioning unit would be automatically taken out-of-service.

## Organization of No. 1 EAX Data Processor Unit

The data processor is the central control unit of the No. 1 EAX. It corresponds in many ways to the central processor frame in the No. 2 ESS. However, it also contains a stored program, in the form of a semipermanent core and drum memory units, which are equivalent to program store in the No. 2 ESS. The data processor is essentially an integrated circuit, general purpose computer whose stored program controls the various call processing and system diagnostic functions. The resultant commands necessary to effect the required system operations are sent to the peripheral units or subsystems by means of input-output devices in the data processor unit. These input-output devices include the communication registers, which are used with the originating and terminating markers, the input-output service buffer connection for the teletypewriter, and the controllers for the magnetic and paper tape units.

The principal components and organization of the data processor unit are represented in the block diagram of Fig. 8-19. The central data processor unit (CPU) with its ferrite-core main memory, the revolving drum memory, and their respective main and drum memory controls make up the computer complex. The register-senders, the maintenance control center, and a number of fault interrupt and related circuits are directly associated with the data processor unit for the transmission of information and the reception of operational instructions. The communications links connected with the computer channel multiplexor provide the data processor with information regarding

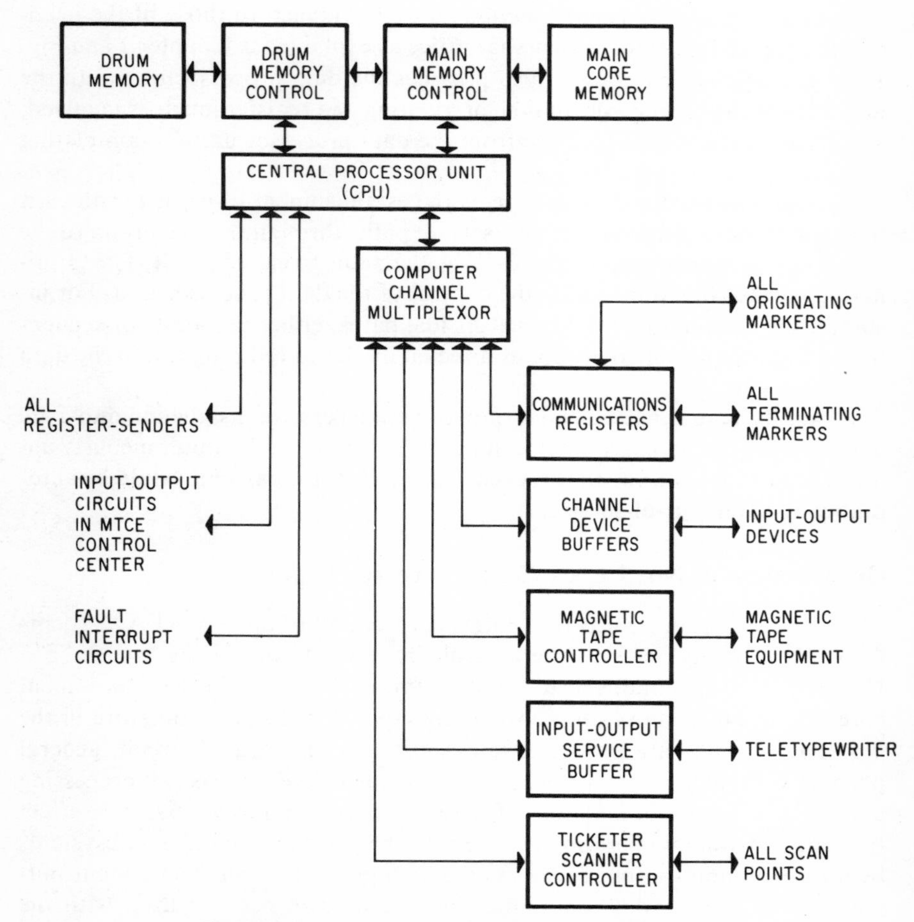


**Fig. 8-19.** No. 1 EAX Data Processor Unit Organization

the processing of calls by the subsystems, status of the subsystems, and data concerning the input-output devices.

The input-output devices connected to the computer channel multiplexor include the communication registers which link the originating and terminating markers, the teletypewriters, and the paper tape equipment. Two other input-output devices, the ticketer scanner and magnetic tape unit, work in conjunction with the stored program to provide automatic message accounting (AMA) if this option is desired.

The stored program, commonly known as software, is contained in the drum memory and the ferrite-core main memory of the data processor unit. This generic program consists of 197,000 binary words, each of 27 bits. The drum memory contains 160,000 binary words and the remainder (37,000 words) are in the ferrite-core main memory. More than half of the stored program instructions are concerned with routine system maintenance and the diagnostics of faults. The information stored in the main and drum memory units depends on the number of lines and trunks equipped in the particular No. 1 EAX office.

Four major programs, each of which is capable of performing one or more system functions, make up the software in the No. 1 EAX. These programs are designated: Operating System, Telephone Applications, Diagnostics, and Office Administration.

The *operating system programs* direct the detailed interface operations required by the other programs for control of the various input-output devices and subsystems. They may be compared to the executive program in the No. 1 ESS because it also schedules the tasks to be performed by the data processor unit. In addition, operating system programs provide many common functions, such as the allocation of buffers and the timing of operations.

*Telephone application programs* contain the control logic and data processing needed to furnish the basic services and features of an electronic switching central office. These programs cover call-processing, ticketing, and metering groups. The call-processing program directs the various functions for processing a call within the No. 1 EAX. For example, it determines the class of service, analyzes the dialed digits, initiates path and route selections, and decides call terminations. The drum memory is used for translations and other data retrievals related to the particular call in progress. The metering program records various system conditions detected by the call-processing programs. Automatic message accounting (AMA) operations are directed by the ticketing programs, which also work in conjunction with the call-processing programs. Ticketing programs determine the answer and disconnect time of calls that are to be ticketed for charge purposes, so that this information can be recorded by the magnetic tape unit.

*Diagnostic programs,* as their name implies, are concerned with system diagnostics and maintenance operations. For instance, errors and faults detected by the various subsystem circuits are conveyed to the diagnostic programs by fault interrupts. These programs endeavor to locate and isolate the faulty unit, and to identify and localize the fault condition. The type of fault and its location are then communicated to the maintenance personnel by means of the teletypewriter in the maintenance control center. Diagnostic programs also provide for routine checks of line, trunk, and junctor equipments.

*Office administration programs* make major and minor changes to the stored program. They also serve to add, delete, or modify data used in the processing of a call.

## Questions

1. What recent technological developments are expected to substantially reduce the cost and size of electronic switching systems? To what part of the electronic switching system would this development apply?
2. Name two or more major electronic telephone switching systems, in addition to the No. 1 ESS, and indicate their manufacturers.
3. What comprises the control complex of the No. 2 ESS? What common unit is associated with it? What equipment unit may be considered equivalent to central control in the No. 1 ESS?
4. Name the major equipment frames in the No. 2 ESS.
5. Compare the semipermanent and temporary memory devices used in the No. 2 ESS with those in the No. 1 ESS. How many bits are in a binary word in the semipermanent memory of the No. 2 ESS?
6. How does the switching network of the No. 2 ESS differ from that of the No. 1 ESS?
7. What are the principal functions of the maintenance center in the No. 2 ESS?
8. What type of power plant is provided for the No. 2 ESS? What is the voltage and phase of the commercial ac input source?
9. What three major equipment groups make up the No. 1 EAX system? What principal elements are in the common control subsystems?
10. What switching device is used in the No. 1 EAX switching network? How does it compare with the similar device employed in the No. 2 ESS?

# GLOSSARY

*Address:* A combination of bits that identifies a location in a memory device or equipment unit.

*AND-NOT gate:* A switching circuit whose output is low (logic 0) only if all of its inputs are high (logic 1). If any of its inputs are low, the output is high.

*Area code:* The three digits, preceding the seven-digit telephone number, that are used for direct-distance-dialing.

*Balanced circuit:* Telephone circuit in which the two conductors are electrically balanced to each other and to ground.

*Bandwidth:* The difference between the limiting frequencies of a continuous frequency band.

*Binary number system:* A number system employing digits 0 and 1 with the base 2, just as the decimal system uses 10 digits with the base 10.

*Bipolar ferreed:* A device comprising two sealed reed switches, a permanent magnet, and a winding with a semipermanent magnet. When the winding is pulsed with current in one direction, the reed contacts close; when the winding is pulsed in the opposite direction, the reed contacts open.

*Bipolar pulse:* A current or voltage pulse that may be either positive or negative.

*Bit (binary digit):* A binary unit of information represented by one of two possible states, such as: on or off, 0 or 1, high or low potential, magnetized or demagnetized.

*Blocking:* Inability to interconnect two idle terminals in the switching network because one or more of the connecting links are in use for another call.

*Buffer:* An isolating circuit used between high-speed and low-speed circuits, or between high-impedance and low-impedance circuits. In call store, it is a register used to store information until such information can be utilized during call-processing operations.

*Bus:* A group of conductors which provide time-shared communication paths for the transmission of information between equipment units.

*Call store:* The equipment unit, in the central processor element of the Nos. 1 and 2 ESS, that provides temporary memory storage of information pertaining to call-processing and maintenance operations.

*Central control:* The equipment unit in the No. 1 ESS that controls the operation of other units in accordance with instructions received from program store.

*Central pulse distributor:* The equipment unit in the Nos. 1 and 2 ESS used by central control to convey signals to various equipments for high-speed control operations.

*Circuit pack:* A small plug-in unit consisting of a printed wiring board containing various components including solid-state devices to constitute one or more circuits.

*Complement:* A binary number formed from a specified binary number by changing each bit from 0 to 1, or vice versa.

*Conditional transfer:* A program instruction which causes central control either to process the next instruction in sequence or to jump to some other indicated instruction, depending upon the results of some previous operation.

*Correed:* A glass enclosed miniature reed switch used in the switching stages of the No. 1 EAX. It is similar to the ferreed except that it is operated only when current flows through its surrounding winding, and releases when current flow stops.

*Counter:* A circuit used to count pulses, including the binary type.

*Crosspoint:* The operated contacts on a ferreed or correed switch.

*Decoder:* A circuit used to translate input information into a form that can be utilized by the receiving unit.

*Destructive readout:* The process which results in not retaining information in a memory device from which it is read out.

*Enable pulse:* Any current or voltage pulse that enables a circuit to become operative.

*Encoder:* A circuit that codes information into a form suitable for transmission from one equipment unit to another.

*End central office:* The local central office which interconnects customer lines and trunks. It is designated a Class 5 office in the DDD or intertoll network.

*Error:* A transient malfunction which cannot be reproduced by the electronic switching system.

*Exclusive OR:* A logical operation for combining binary words so that, given words A and B, the result is a binary word that has a 1 only in those bit positions in which either A or B, but not both, has a binary 1.

*Fault:* A circuit malfunction that can be reproduced by the electronic switching system.

*Ferreed:* A device consisting of two miniature glass enclosed reed switches (similar to the correed) which are operated or released by controlling the magnetization of two adjacent remendur plates. The magnetization of these plates is controlled by two windings. When both windings are energized, the reeds operate and remain closed. When only one winding is energized, the reeds release or remain open.

*Ferrod:* A current-sensing device used in scanners in the No. 1 ESS for supervisory and related functions.

*Flip-flop:* A solid-state device capable of assuming either of two stable states (set or reset) for storing a binary digit of information. It remains in either state until a signal changes it to the other state.

*Gate:* A basic circuit which produces an output only when certain input conditions are satisfied.

*Grid:* A two-stage switching network in which a single path exists between every first-stage switch and every second-stage switch.

*Hardware:* Equipments, circuits, devices, and other apparatus installed in electronic switching offices.

*Hopper:* An area in a temporary memory unit, such as call store, used to record a list of items for subsequent communications with processing programs or input-output programs sent to central control.

*Index register:* A register used to store a numerical quantity for the modification of data or address information.

*Input-output:* The process of transmitting information from an external source to an equipment unit, or from an equipment unit to an external source.

*Instruction:* A binary word or groups of binary words which direct central control in the No. 1 ESS or equivalent processing control units in other electronic switching systems, to perform a particular function.

*Interoffice trunk:* Telephone channel between two separate central offices.

*Interrogate:* To determine the state of a device or circuit.

*Intraoffice trunk:* Telephone channel used to interconnect two customer lines within the same central office.

*Junctor:* Any path or connecting circuit between switching frames in the switching network.

*Link:* A single transmission path between terminals on one switch and the terminals on the switch in the next switching stage.

*Logic circuit:* A circuit capable of producing one or more outputs only when specified input conditions are satisfied.

*Logical product* (AND): A logical operation for combining binary words so that, given binary words A and B, the result is a binary word that has a binary 1 only in those bit positions in which both A and B have a 1.

*Logical union (Inclusive OR):* A logical operation for combining binary words so that, given binary words A and B, the result is a binary word that has a 1 only in those bit positions in which either A or B have a 1.

*Loop:* The two-wire circuit formed by the customer's telephone set, cable pair, and other conductors that connect it to the central office equipment.

*Magnetic core:* A toroidal device (shaped like a doughnut) that is capable of storing information by remaining magnetized in either of two directions.

*Mask:* A binary word used in masking to specify the bit positions to be changed to 0 (unless they are already 0) in a binary word to be masked.

*Masking:* The process of changing certain bit positions of a binary word to 0 as specified by the mask. The binary word which is to be masked and the word to be used as a mask are combined by means of the logical AND product.

*Memory:* A unit or device into which information, in the form of binary digits (bits), can be placed and extracted at a later time.

*Memory card:* A thin aluminum card with small bar magnets used to retain binary information in program store of the Nos. 1 and 2 ESS.

*Memory circuit:* A circuit which, having been placed in a particular state by an input signal, will remain in that state after the removal of the input signal.

*Module:* An equipment unit capable of being combined with other similar units to form a larger unit.

*Off-hook:* The condition that indicates a closed loop or the active state of a customer's line.

*Office code:* The first three digits of the seven-digit local telephone number (directory number).

*On-hook:* The condition that indicates the idle state or open loop of a customer's line.

*Parity bit:* A binary digit (bit) attached to a binary word to make the total number of 1's, including the parity bit, an odd or even amount.

*Parity check:* A check on the validity of a binary word by determining whether the number of 1's in the word is odd or even.

*Peripheral bus:* A special multipair cable that interconnects major units in the No. 1 ESS and similar electronic switching offices.

*Peripheral units:* Equipment units that connect to the central processor unit. For example, the scanners, signal distributors, network controllers, and the master control center in the No. 1 ESS are peripheral units.

*Program:* An organized set of instructions used to control operations of an electronic switching system.

*Program store:* The semipermanent memory unit in the Nos. 1 and 2 ESS that stores the processing instructions and translating information.

*Queue:* An area in the temporary call store memory used to record a writing list for some particular function. For example, the writing list or queue for customer dial pulse receiver circuits.

*Random access:* The ability to gain access to any location of a memory unit in a time that is independent of the location itself.

*Readout:* The information extracted from a memory device such as program store or call store.

*Real time:* The actual time of occurrence of an event.

*Redundancy:* The use of additional equipment and facilities to provide for continuity of service during trouble situations.

*Register:* A binary word repository or device for temporarily storing binary words in control units.

*Reset:* To restore a storage device, such as a flip-flop, to a prescribed state which is opposite to the set state.

*Ringing current:* 20 Hz ac at a voltage of 75-105 volts supplied by the central office to ring a customer's telephone bell.

*Rotate:* The process by which each bit of a word is moved either to the right or to the left by a specified number of bit positions. Bits passing through the end of the word reenter at the opposite end of the word.

*Scanner:* The equipment unit in the No. 1 ESS that provides central control with information access to lines, trunks, and test points.

*Scanpoint:* The place where a ferrod is connected in order to determine the state of a customer's line, trunk, or a test point.

*Semipermanent memory:* A read-only memory containing information which cannot be changed by the internal circuitry of the system. Changes must be made by external means.

*Service circuit:* An interconnecting circuit in the switching network that may be connected to lines or trunks as required to perform various functions, such as dial pulse detection and audible ringing.

*Set:* To place a storage device in a prescribed state which is opposite to the reset state.

*Shift:* The process by which each bit of a binary word is moved either to the right or to the left by a specified number of bit positions. Bits that pass through the end of the word are not retained. Vacated bit positions are filled with zeros.

*Signal distributor:* The equipment unit in the No. 1 ESS that provides access from central control to relays in trunk and junctor circuits by acting as a buffer.

*Signal processor:* The equipment unit in the No. 1 ESS used in the larger electronic offices to perform repetitive input-output tasks for central control.

*Software:* Instructions, translations, and related data in the form of binary words which constitute the stored program in an electronic switching system.

*Subroutine:* A sequence of programmed instructions to perform a particular function that is common to several programs.

*Talking path:* The transmission path of a telephone circuit making up the tip and ring conductors, and the equipments connected to them.

*Temporary memory:* A read and write memory containing information that can be changed by the internal circuitry of the electronic switching system.

*Time-division-multiplex:* The process of transmitting two or more signals over a common path by using different time intervals for each signal.

*Translation information:* Information contained in program store pertaining to individual lines and trunks. It may be used, for example, to convert an equipment location into its related directory number or to ascertain the class of service of a customer's line.

*Translator:* A circuit used to convert information from one form to another.

*Unconditional transfer:* An instruction which causes central control to proceed immediately to a particular instruction.

*Unipolar pulse:* A pulse having only one polarity.

*Word (binary word):* A set of characters, such as binary digits (0 and 1), used to express information in an electronic switching system or computer.

*Write:* To insert information into a memory device.

# ABBREVIATIONS USED IN THE TEXT

| | |
|---|---|
| AMA | automatic message accounting |
| ANC | all-number calling (plan) |
| AOL | add-one logic |
| AOR | add-one register |
| ASW | all seems well |
| BCD | binary-coded decimal |
| BOWR | buffer order word register |
| CO | cut-off |
| CPD | central pulse distributor |
| dB | decibel |
| DBR | data buffer register |
| DDD | direct-distance-dialing |
| DPU | data processor unit |
| EAS | extensive area service |
| EAX | electronic automatic exchange |
| ESS | electronic switching system |
| FCG | false cross and ground |
| Hz | Hertz |
| IC | integrated circuit |
| I/O | input/output |
| MF | multifrequency |
| OGT | outgoing trunk |
| OWD | order word decoder |
| OWR | order word register |
| PABX | private automatic branch exchange |
| PAR | program store address register |
| PBX | private branch exchange |
| PCM | pulse code modulation |
| POB | peripheral order buffer |
| PU | peripheral unit |
| S, C | sleeve, control |
| SP | signal processor |
| S, R | set, reset |
| S × S | step-by-step |
| TDM | time division multiplex |
| T, R | tip, ring |
| TTY | teletypewriter unit |

# INDEX